GEDDES & GROSSET

DICTIONARY OF GEOGRAPHY

This edition published 2001 by Geddes & Grosset,
an imprint of Children's Leisure Products Limited

First published 1998
Reprinted 2001

David Dale House, New Lanark Scotland, ML11 9DJ

ISBN 1 85534 336 3

Printed and bound in the UK

Contents

A

ablation the removal of surface snow or ice by melting or evaporation. When ablation occurs at the edges of glaciers, large quantities of DEBRIS are deposited to form *ablation moraines*.

abrasion effect of the wearing away of rock by fragments of other rock carried by moving ice, water or wind. An *abrasion platform* is a rock platform extending from the foot of a sea cliff.

absolute humidity the amount of water vapour in the air, expressed in grams per cubic centimetre. Air at a given temperature and pressure is capable of holding a specific amount of water vapour, after which it becomes saturated and DEW POINT is reached. Cold air has a lower absolute humidity than warm air at the same pressure.

absolute temperature the Kelvin temperature scale based on ABSOLUTE ZERO.

absolute zero the lowest temperature theoretically obtainable, 0K = –273.15°C (–459.67°F).

abyssal the bottom zone of the ocean, below 3000 metres (9842 feet), where light does not penetrate and temperatures are below 4°C (39°F). *Abyssal plains* are wide, flat areas that result from the accumulation of fine sediments, up to a kilometre (3280 feet) thick, on top of the ocean floor crust.

accordant drainage a drainage pattern that is related to the structure of the underlying rocks. If the main stream flows in the same direction as the dip of the rock, it is said to be *ac-*

cordant. If the main stream bears little or no relationship to the underlying geology it is DISCORDANT.

accordant junctions the law of this states that tributaries join a stream or RIVER at the same elevation as that of the larger watercourse. There is no sudden drop in the level of the tributary at the junction.

accordant summits hill or mountain tops at approximately the same height above sea level, thought to be evidence of either uplift and subsequent erosion or a region where surface material has been washed away leaving hill tops at a uniform level.

accretion (1) the growth of land by the offshore depositions of a sediment. SPITS are formed by accretion. (2) in meteorology, hailstones grow by accretion as more water freezes onto them.

accumulation zone that part of a slope that, over a period of time, gains material, leading to a progressive raising of the ground surface.

acid rain rain with a high concentration of pollutants such as dissolved oxides of sulphur (mostly from burning coal) or oxides of nitrogen (mostly from vehicle exhausts). Acid rain is harmful to plants and animals and is responsible for the destruction of forests in parts of Europe.

acid soil a soil with a pH of less than 7. Acidity may be due to the nature of the rock and vegetation from which the soil was produced or to the leaching out of soluble salts, mainly of sodium and calcium.

active glacier a glacier that is receiving new ice above the snow line but is losing ice by ABLATION below the snow line.

active remote-sensing system a remote sensing system having its own source of electromagnetic radiation (*see* ELECTROMAGNETIC WAVES), as is the case for RADAR sensors. This system measures the electromagnetic radiation produced by the sensor, which has been reflected from the ground surface.

adiabatic temperature change a temperature change that involves no heat transfer but is due to a change in pressure; e.g. as air rises, pressure decreases, volume increases and temperature falls.

adobe (1) bricks of sun-dried clay. (2) clayey and silty deposits in Mexico and southwest USA.

adret the sunny side of a hill or valley.

adsorption the taking up of one substance at the surface of another; e.g. mineral particles in the soil adsorb dissolved substances with which they come into contact.

advection the horizontal transfer of heat in gases and liquids. *Compare with* CONVECTION. It is the process by which heat is moved from tropical areas (*see* TROPICS) towards the POLAR areas, either in the ATMOSPHERE when AIR MASSES move or in OCEAN CURRENTS such as the GULF STREAM.

aeolian pertaining to the wind. *Aeolian erosion* leads to the wearing away of rock surfaces, prior to the wind carrying away the abraded material, which may be deposited as DUNES or LOESS.

aerosol a fine mist or fog of particles in a gas. Aerosols enter the ATMOSPHERE from natural sources, such as volcanoes, or from human activities, such as burning FOSSIL FUELS. They lower the amount of solar radiation reaching the Earth. thus cooling the surface. In meteorology, aerosols may include dust, sea salt and organic matter in addition to smoke.

aftershock the relatively minor tremors following an earthquake. Aftershocks may continue for hours, days or months, depending on the nature of the original earthquake.

agglomerate a rock of volcanic origin made up of fine fragments cemented together in a fine matrix.

aggradation the building up of the surface of land by waterborne or wind-borne material. Rivers deposit material if their velocity of flow decreases or if there is an increase in sediment. Building a dam across a river can cause aggradation. The accumulation of material in forming sea beaches is also aggradation.

agonic line a line drawn on a map joining the north and south magnetic poles.

agronomy the theory and practice of crop production and the management of soils.

Aghulhas current a warm ocean current off the coast of south-east Africa.

air frost air with a temperature at or below 0°C (32°F).

air mass an area of the ATMOSPHERE that, horizontally, has more or less uniform temperature and humidity and extends for hundreds of kilometres. Air masses acquire their properties from prolonged contact with their areas of origin. Air masses are separated by fronts. Depending on the area of origin, an air mass may be Antarctic, Arctic, polar, equatorial or tropical. These are abbreviated as AA, A, P, E and T respectively. The air mass may then be continental (c), maritime (m) or monsoonal (M), depending on its humidity. Further categories based on stability and movement allow a complete categorization.

air stream a current of air, a wind, blowing from an identifiable source.

ait *or* **eyot** a small island in a river. The abbreviated form *ey* survives in many British place names, denoting a former island, e.g. Romsey.

alas a large, steep-sided, flat-bottomed depression occurring in a PERMAFROST area. It may be several kilometres across and often contains lakes.

alimentation the snowfall, avalanche, snow and ice, together with any re-frozen meltwater, that accumulates on a glacier.

alkaline soil a soil having a pH greater than 7 and usually indicating a high concentration of carbonates, mainly of sodium or calcium.

alluvium deposits produced as a result of the action of streams or RIVERS. Moving water carries sediment, particles of sand, mud and silt, and the faster it moves the greater the load it can carry. When the velocity of the water is checked by it meeting a stationary or slower moving body of water, then much or all of the sediment is dropped, forming alluvium. In mountainous regions where streams descend to lowlands, a fan or cone of sediment may build up. Rivers in areas of less extreme geography deposit alluvium on the flood plain where the water changes velocity along the curving course of the river, or when rivers overflow their banks during floods.

alp a shoulder of land in mountainous terrain standing above a GLACIATED TROUGH and stretching up to the summer snow line. It is snow-covered in winter but provides good grazing in summer.

altimeter an instrument for indicating height above sea level.

altitude (1) the height above a given level, e.g. sea level. (2) in surveying, the angle between the horizontal and a point at a higher level.

alto a term referring to clouds between 3000 and 6000 metres (9840 and 19,684 feet) high, as in *altocumulus*, a tall cloud formation, or *altostratus*, a grey sheet of cloud.

anabatic wind an upslope wind formed when air on the sides of a valley is heated more quickly than air above the valley floor. The heated air rises and is replaced by cooler air from the valley floor.

ana-front a FRONT where warm air rides over colder air.

anaglyph a method of obtaining a three-dimensional image of topography by viewing two adjoining aerial photographs that have been printed in red and green by using special lenses, one that is tinted red and the other green.

anastomosis the division of a river into two or more channels with large islands in between.

anemometer an instrument that measures wind speed. A *vane anemometers* consists of three or four conical cups mounted on arms on a vertical spindle. As the cups are blown round, the spindle drives a generator. The faster the spindle rotates, the greater the output of the generator, which is recorded on paper. A *pressure-tube anemometer* is kept facing the wind by a wind vane. The wind pressure down the tube is converted to a reading.

aneroid barometer *see* BAROMETER.

angle of declination the angle between true north and the direction of the magnetic MERIDIAN.

Antarctic regions south of the Antarctic circle, 66.6°S.

Antarctica the land mass within the Antarctic circle. All but 5 per cent of its area lies under ice that can reach 4000 metres (13,000 feet) in some places. There are no freeze-thaw seasons, so that exposed rocks are little affected by weathering. Much of the surface is featureless, with huge GLACIERS and CREVASSES. The contrast between the icy land mass and the warmer ocean produces a strong temperature gradient along the polar FRONT and strong west winds.

Antarctic convergence the zone in the seas around Antarctica where cold, heavy seas sink below warmer waters to the north.

anticyclone a region of relatively high atmospheric pressure measuring thousands of kilometres across. It is also known as a high. ISOBARS are widely spaced, indicating either light winds or calms. The weather associated with anticyclones is settled, being warm, sunny and dry in summer and clear and frosty or foggy in winter. The pressure is usually 1000 millibars and above, and air movement is clockwise in the northern hemisphere and anticlockwise in the southern.

antimeridian any MERIDIAN that is 180° of longitude from any other meridian.

antipodes points on the Earth's surface that are directly opposite each other. For example, the Antipodean Islands, southwest of New Zealand, are at the antipodes of the Channel Islands.

aphelion the point on the orbit of a planet when it is at its farthest distance from its sun. On 4 July the Earth is at aphelion, 152 million kilometres (94.5 million miles) from the Sun.

aphotic zone a zone deeper than 300 metres (635 feet) in lakes and oceans where light does not penetrate and photosynthesis, the process whereby plants turn the Sun's energy into food, is impossible.

apogee (1) the point on the orbit of a moon or satellite where it is farthest away from the planet or point around which it is circling. When our moon is at apogee, its effect on the tides is at its smallest and the difference between high and low tide is reduced. Another reference concerns the meridional altitude of the Sun on the longest day. (2) the highest point reached by a rocket, used in remote sensing, when fired upwards from the Earth.

apparent time local time, established from the time when the Sun is at its highest.

Appleton layer the upper layer of the IONOSPHERE, which is approximately 300 kilometres above the Earth's surface. It reflects shortwave radio waves back to the Earth.

aquifer a layer of rock, sand or gravel that is porous and therefore allows the passage and collection of water. If the layer has sufficient porosity and permeability it may provide enough GROUNDWATER to produce springs or wells. If the layers above and below the aquifer are impermeable, the water is under pressure (*hydrostatic pressure*) and can be extracted in an artesian well whilst the level of the well is lower than that of the WATER TABLE. A significant proportion of London's water supply comes from the London Basin (an ARTESIAN BASIN), although the supply is diminishing because the water table and pressure have fallen because of prolonged extraction of water. In many cases pumps may be required to raise the water to the surface.

arable land land that is suitable for growing crops.

arch an opening on the sea coast where two caves, originally on opposite sides of a headland, have been eroded until they meet. Eventually the roof of the arch will collapse and the seaward end becomes a STACK.

archipelago a cluster of islands.

Arctic regions north of the Arctic circle, 66.6°N.

Arctic front a FRONT lying between 50°N and 60°N where cold Arctic air meets slightly warmer polar air. As the temperature difference between the two air masses is slight, the front is not very active. It often extends from south Greenland to north Norway.

Arctic smoke fog formed in high latitudes when cold air condenses as it passes over a warmer water surface.

arena a shallow, usually circular, basin surrounded by higher land.

arête a French word used to describe the ridge between two CIRQUES.

aretic an area that does not have flowing streams. An example is hot deserts, where any PRECIPITATION either soaks into the ground or evaporates.

arid this means dry. An *arid area* has less than 250 millimetres of PRECIPITATION annually, which is insufficient to support vegetation. An *arid zone* may also be defined as one where the potential for EVAPORATION exceeds precipitation.

artesian basin when permeable rocks are folded into a syncline and water seeps into this AQUIFER (formed from the permeable strata) until the rock becomes saturated and the water is then under pressure. If a borehole is sunk into the aquifer, the pressure on the water will drive it to the surface without the need for pumping. This is an *artesian well*.

an artesian basin

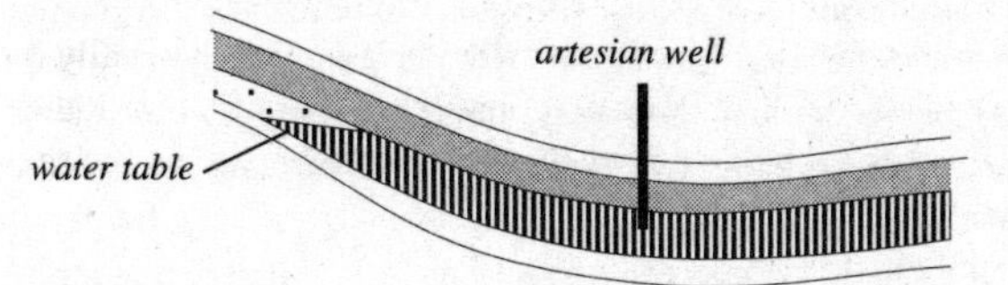

aseismic plates any of the Earth's crustal plates where there are few earthquakes (*see* PLATE TECTONICS).

aspect the direction in which a slope faces.

asphalt a naturally occurring tar. Most asphalts are thick and viscous and may be almost solid, but some can be poured without being heated. They are often formed as a residue when the lighter fraction of an oil pool has evaporated. Two of the best-known examples are the Pitch Lake in Trinidad and the Athabasca Tar Sands in Canada.

asthenosphere the higher layer of the upper mantle of the Earth, which, in comparison to the LITHOSPHERE upon which it rests, is relatively weak. The zone that accommodates lateral, plastic flow is partly molten and is the site for MAGMA generation.

astrobleme an ancient, crater-like feature on the Earth's surface, thought to have been caused by collision with an extraterrestrial body.

Atlantic-type coast where the ridges and valleys determined by the underlying geology run transversely to the coast. If the sea floods the coastal lowland, RIAS or FJORDS may result.

atmosphere the layer of gases and dust surrounding the Earth, which can be divided into shells, the lowermost being the TROPOSPHERE, which is overlain by the STRATOSPHERE. The density falls with height, and because it is thinner, breathing is more difficult at high altitudes. Almost 75 per cent of the total mass is contained within the troposphere. The main gases found in the atmosphere are nitrogen (78 per cent), oxygen (21 per cent), the inert gas argon (0.9 per cent), carbon dioxide (0.03 per cent) and then very small amounts of other inert gases, with methane, hydrogen, water vapour and ozone. The ozone exists mainly in a layer (*see* OZONE LAYER) at a height of about 25 to 30 kilometres (15 to 19 miles), although it is found elsewhere within the stratosphere.

atoll a circular or horseshoe-shaped coral reef surrounding a LAGOON. Atolls are particularly common in the Pacific Ocean. The outer rim lies in shallower water, while the inner edge shelves steeply to the deeper water of the lagoon.

aurora veils, sheets and rays of white or coloured light seen flickering in the night sky. It is thought to be caused by the ionization of atmospheric molecules by solar and cosmic radiation as the effect is more likely to occur during periods of high sunspot activity. The phenomenon is especially marked in polar latitudes and known as *Aurora Borealis* or *Northern Lights* in the northern hemisphere and *Aurora Australis* or *Southern Lights* in the southern hemisphere.

avalanche a rapidly descending mass of snow down a mountain-side. Wet snow avalanches are the most powerful, cause most damage and can even erode the bedrock over which they move. They are most likely to occur in spring when melting snow produces large quantities of water. Various factors are used to classify avalanches: snow humidity; type of breakaway, whether loose snow or a slab; position of the sliding surface; movement type, whether a flow on the ground or powder carried in the air.

avalanche wind the blast of air that moves before an AVALANCHE. It can be very destructive, causing buildings to explode before the avalanche actually reaches them.

aven a vertical opening into a limestone cave.

axis the Earth's axis is the line joining the north and south poles, around which the Earth rotates every 24 hours.

azimuth in surveying, the horizontal angle, measured clockwise, from true north (*true azimuth*) or magnetic north (*magnetic azimuth*) to another point. For example, the azimuth of a point due east is 90°.

azonal a term that refers to a young soil developing on a bare rock surface, e.g. SAND DUNES and ALLUVIUMS.

B

background level the naturally occurring level of radiation activity. This varies from area to area according to the type of rock in the ground (e.g. granite is one of the more radioactive rocks), and also to the amount of sunshine received, amongst other factors. Background level can also refer, in some circumstances, to levels of pollution, e.g. noise pollution or light pollution.

backing a change of direction of wind, anticlockwise in the northern hemisphere and clockwise in the southern hemisphere. Thus in the northern hemisphere a change of direction from northerly to westerly would be backing. (*See also* VEERING).

backshore that part of a beach between high water during normal spring tides and the foot of the cliff.

back slope the slope of a CUESTA, which is more gentle and follows the DIP of the STRATA.

back wall the steep rock wall at the back of a glacially eroded hollow or CIRQUE.

backwash the return flow of water to the sea after a wave has broken. Short, steep waves break vertically onto a beach, causing a strong backwash that drags beach material seaward.

badlands country that is difficult to travel across because of frequent deep RAVINES and GULLIES and steep, sharp ridges. The landscape is usually created by fluvial erosion in a semi-arid area. Overgrazing can encourage soil erosion and add to the problem.

bank (1) a hillside. (2) a muddy, sandy or shelly deposit in shallow sea water. Inshore, banks are in very shallow water. Offshore, the term is often applied to a fishing ground that is shallower than the surrounding sea, e.g the Dogger Bank. (3) the margins of a RIVER channel.

bar (1) a deposit of sand or mud in a RIVER channel and often occurring across the mouth of a river or across the entrance to a harbour. (2) sand, shingle or mud deposited in a long deposit in the sea, often roughly parallel to the coastline. (3) a unit of atmospheric pressure. Atmospheric pressure is measured in millibars or thousandths of a bar.

barbed drainage a pattern of drainage in which TRIBUTARIES meet the main river at an obtuse angle as though they try to flow upstream. The pattern is thought to result from RIVER CAPTURE, which, in effect, reverses the flow of the main river (*see* DRAINAGE for diagram).

barometer an instrument for measuring atmospheric pressure (*see* BAR). An *aneroid barometer* is a thin, corrugated metal box containing a partial vacuum. The sides move in response to rising or falling pressure, and this movement is magnified and transmitted by a series of levers to a needle on a dial. The atmospheric pressure can then be read. A *barogram* is an aneroid barometer connected to a pen, which records changes in atmospheric pressure on a moving cylinder of paper.

barrier beach an elongated sand or shingle BANK that is not submerged at high tide. When it is high enough for DUNES to form it is known as a *barrier island.*

barrier lake any lake formed by a naturally occurring barrier of, for example, rock, LAVA, ice MORAINE, a landslide, etc, that interrupts the normal flow of a river.

barrier reef *see* CORAL REEF.

barrow a communal burial mound dating from the Stone Age until Saxon times.

baseline a line on the ground that is surveyed extremely accurately and from which other readings are taken during a survey. It is claimed that a base line is surveyed to an accuracy of 1 in 300,000.

basalt a dark, fine-grained igneous ROCK containing feldspar (plagioclase), pyroxene and possibly olivine. Basalt occurs as LAVAS (extrusion) and minor intrusions. Basalt flows cover a vast proportion of the Earth's surface (over two-thirds) and the 'terrestrial' planets (Mercury, Venus and Mars). Two subdivisions exist—alkali basalts and tholeiites—the former being found in regions of rifting, crustal deformation and on oceanic islands, while the latter typify ocean floor and stable continental crust extrusion.

basic rocks quartz-free IGNEOUS ROCKS that have a low silica content but a high proportion of magnesium iron and calcium.

BASALT and gabbro are typical examples. They are also classified on the composition of the mineral feldspar and whether it has more calcium or sodium.

basin (1) a depression in the land surface. (2) a huge depression occupied by sea water, e.g. an ocean basin. (3) the catchment area of a river system. (4) a geological feature, such as a large depression filling with sediment or a sinking caused by solution and removal of mineral deposits (e.g. salt).

battue ice large ice floes that obstruct the St Lawrence estuary in winter and pose a considerable danger to shipping.

bauxite the primary ore source for aluminium, formed by the weathering of aluminium-bearing rocks in tropical conditions. A residual deposit that, if it contains more than 25 per cent aluminium oxide, can be exploited commercially.

bay an open, curving recess in the coastline of a SEA or LAKE, e.g. Cardigan Bay.

bayou an area of slow-moving, marshy (*see* MARSH) water or a cut-off MEANDER that forms a backwater beside the main channel. The word was originally applied by French settlers to the bayoux in the Mississippi DELTA.

beach budget in most beaches, the material lost by EROSION or BACKWASH is compensated for by deposits carried from other areas, e.g. by a RIVER or by LONGSHORE DRIFT, or it may be replaced by SWASH. It is important that the balance of this budget is not lost, as beaches protect coastlines from erosion. This should be taken into account when designing breakwaters, etc, that might disturb the budget.

bearing the horizontal angle between a BASELINE and the point being surveyed. In navigation, bearings are measured in degrees, clockwise from the north.

Beaufort scale a system for indicating wind strength, developed in 1805 by Admiral Sir Francis Beaufort. Measurements are taken 10 metres (33 feet) above the ground, and each of 12 levels is characterized by particular features on the landscape or certain effects upon people or objects.

The Beaufort scale

	Velocity of wind, metres per second	*Descriptive weather condition*	*Identifying features*		
			at sea, wave height (m)		*on land*
0	< 0.3	calm	0	mirror-like	smoke rises vertically
1	0.3–1.5	light air	0.2	ripples	smoke indicates wind direction
2	1.6–3.3	light breeze	0.4	wavelets	wind vane moves; wind felt on face
3	3.4–5.4	gentle breeze	0.8	wavelets with breaking crests	twigs move; light flags extended
4	5.5–7.9	moderate breeze	1.5	small waves	small branches move; dust and loose paper raised
5	8.0–10.7	fresh breeze	2.0	moderate waves, white horses	small leafy trees sway; wavelets on inland water
6	10.8–13.8	strong breeze	3.5	large waves, foam crests	large branches move; whistling in telegraph wires
7	13.9–17.1	near gale	5.0	breaking waves	whole trees in motion; some resistance from wind when walking
8	17.2–20.7	gale/fresh gale	7.5	moderate high waves	twigs broken off trees; resistance slows walking
9	20.8–24.4	strong gale	9.5	high waves, toppling crests	chimney pots and slates may be removed
10	24.5–28.4	storm	12	very high waves, overhanging crests	considerable structural damage; trees uprooted

The Beaufort scale contd

	Velocity of wind, metres per second	*Descriptive weather condition*	*Identifying features*		
			at sea, wave height (m)		*on land*
11	28.5–32.7	violent storm	15.0	extremely high waves; ships lost to view	widespread damage
12	> 32.7	hurricane	> 15	sea all white; foam/sprayed filled air	catastrophic damage; heavy loss of life

beck the north of England term for a rapidly flowing stream.

bed (1) the floor of a RIVER, lake or sea. (2) a layer of rock within a sedimentary rock sequence. It is a discrete layer that differs from the rest of the sequence because of a difference in composition, texture or structure.

bedrock the unweathered rock below the soil.

ben the Scottish Gaelic term for a MOUNTAIN or peak, e.g. Ben Nevis.

benchmark a permanent point of reference used by a surveyor. It may be marked on a weather-resistant rock or on a stone building. In Britain, it gives the height of the benchmark above the ORDNANCE DATUM, correct to one place of decimals.

Benguela current a cold OCEAN CURRENT flowing northwards off the west coast of South Africa. It has a cooling effect on the coastal CLIMATE of the area as well as bringing nutrients to the fish shoals, so making a lucrative fishing industry possible.

Benioff zone the EARTHQUAKE zone where one continental plate is pushed under another when they meet (*see also* PLATE TECTONICS).

benthic a term describing a plant or animal living on the bottom of a LAKE or sea. This may include crawling or burrowing at the sediment-water boundary, being attached to it (as with seaweeds), or purely sessile.

berg (1) the German word for 'mountain', which occurs in the names of mountains, e.g. *Vogelsberg*, or mountain ranges, e.g. the *Drakensburg*. (2) a colloquial word for an iceberg.

bergschrund a deep crevasse at the back of a CIRQUE GLACIER, marking the line where the glacier ice is falling downslope away from the back wall of the cirque.

berg wind a hot, dry wind that blows down from the high interior of South Africa towards the east coast, causing oppressive conditions. It occurs when there is a DEPRESSION to the west of the high ground and is similar to a FÖHN wind and CHINOOK.

berm a high ridge of shingle on a beach marking the limit of SWASH.

bill a beak-like headland, e.g. Portland Bill.

billabong an Australian Aborigine term to describe a temporary stream or cut-off pool in an abandoned MEANDER.

bioclimatology the study of CLIMATE and its effect on living things, especially in relation to their health. (*See windchill* in WIND).

biogeosphere the upper crust and surface of the Earth, which contains organic life.

biomass the total mass of all the living organisms in a given area that can be supported at each level in the food chain. It is expressed as mass per unit area. The *standing crop* is the amount of living material present in an area and usually diminishes with each step in the food chain away from plants. Aerial photography is used to estimate the biomass of the oceans. Knowing the biomass of an area of sea is important for conserving fish stocks.

biosphere the zone where life exists, extending from roughly 3 metres (10 feet) below ground surface to approximately 30 metres (100 feet) above it. It also includes both sea and fresh water to a depth of 200 metres (650 feet).

biota all the plants and animals of a given area, i.e. the flora and fauna of a particular ecosystem.

biotic relating to life or living things (hence BIOTA). Thus, for an organism, the other living organisms around it comprise the *biotic environment* and may be competitors, predators, parasites, etc.

blanket bog a continuous covering of PEAT BOG formed where there is high rainfall and acid soil, e.g. in Ireland or northern Scotland (*see* BOG; PEAT).

blind valley a steep-sided valley with a cliff across the lower end. Any stream usually goes underground as it approaches the lower end of the valley. Blind valleys may have been formed by roof collapse, which exposes an underground stream.

blizzard a strong wind, carrying heavy snow, which may lead to a WHITE-OUT.

blowhole a hole in the top of a cliff through which air and possibly sea water blow as a result of waves surging in from and out to sea through an underground tunnel in the cliff.

blowouts DEFLATION HOLLOWS in sand DUNES that lack vegetation cover. They are caused by wind EROSION, vary in size from a metre or so across to several kilometres, and are usually temporary features. The windblown material removed from the blowout usually collects elsewhere as a dune, sometimes called a *blowout dune*.

blue-band a band of bluer ice in a GLACIER. It is bluer because it is bubble-free and denser than the surrounding ice. It may be due to meltwater in a CREVASSE refreezing.

bluff (1) a high or steep HEADLAND. (2) a steep slope on the outside of a MEANDER.

bocage a landscape comprising small fields and low hedges, typical of Brittany and Normandy. Inevitably, bocage is disappearing as fields are enlarged to render farming more economical.

bog an area of waterlogged, spongy ground, made up primarily of decaying vegetation, especially rough grass, rushes and sphagnum moss. It covers areas of Russia, Canada and Scandinavia (*see* BLANKET BOG, PEAT) and is often the site of surprisingly good preservation of, for example, mammoths.

bora a strong, cold, dry winter wind blowing down from the Balkan MOUNTAINS to the Adriatic Sea. It occurs when a deep depression passes over the Mediterranean Sea while there is a high pressure area over central Europe. The wind tends to be channelled by the surrounding mountains.

bore the incoming tidal flow moving swiftly up a river as a moving wall of water. The height of a bore gradually lessens as it moves upstream. The most notable bore in Britain is on the River Severn, but bores also occur on the Trent and Yorkshire Ouse.

Boreal a climatic zone typified by short summers and long winters that are cold and snowy. An early post-glacial era within Europe that has been dated at 7500–5500 BC and was drier than the present climate. The forests became mixed oak, hazel and pine.

boreal forest the CONIFEROUS FOREST of the TEMPERATE latitudes of the northern hemisphere.

borehole a hole drilled during prospecting for coal, oil or natural gas to enable samples to be taken.

boss a roughly circular, steep-sided outcrop of intrusive igneous rock that is usually a few square kilometres in area.

boulder a rounded piece of rock more than 200 millimetres in diameter.

boulder clay a deposit that is glacial in origin and made up of boulders of varying sizes in finer-grained material, mainly CLAY. It is laid down beneath a GLACIER or ICE SHEET and shows little or no structure. Large blocks plucked from the terrain over which the ice has moved may be found in a matrix of finer material that has been ground down by the glacier. An alternative and more frequently used term is *till*, of which several types have been defined, depending on their specific mode of formation and position within the ice body. *Moraine* is an associated term referring to ridges of ROCK debris carried and deposited by ice sheets or glaciers of the various types; *lateral moraine* accumulates at the edge of a glacier and *terminal moraine* at the leading edge. *Ground moraine* is the same as boulder clay.

boundary current the currents flowing at depth in the ocean where there is a marked difference in temperature and salinity between the current and the rest of the sea.

bourne a temporary steam, after heavy rainfall, in chalk country. Rainwater percolates down through the chalk until it reaches the WATER TABLE. In time, the ground above becomes sufficiently saturated to allow a stream to flow in an otherwise dry valley.

brackish a term for slightly salty water containing between 15 and 30 parts of salt per 1000 parts of water. It is intermediate between sea and fresh water.

braided channel a RIVER channel that has deposited bars and islands. Braiding occurs where the river becomes shallow, the rate of flow slows and the river cannot carry its full load of sediment. It may also occur where the banks are easily eroded.

breaker a very steep wave that breaks with a pronounced SWASH onto the shoreline. *Plunging breakers*, which curl over and break with a crash, are destructive. *Spilling breakers*, which break gradually over some distance, and *surging breakers*, which surge up the beach, are constructive and deposit material on the beach.

breccia a sedimentary rock comprising coarse angular clasts (rock fragments). The nature of the rock (i.e. angular fragments reflecting lack of weathering before deposition) implies deposition very close to the source area.

breckland (1) this is used generally when referring to heathland. (2) an area of poor, sandy, gravelly soil in East Anglia that has been planted with conifers by the Forestry Commission.

brig a hard, rocky, coastal HEADLAND, e.g. Filey Brig in Yorkshire.

broads a series of shallow, reed-fringed lakes linked by a slow flowing river, e.g. in East Anglia. It is thought that the Norfolk and Suffolk Broads may have been created by PEAT removal in medieval times, leaving shallow depressions that were later flooded.

brow the upper part of a hill or mountain where the steep slope becomes gentler.

brown earth a wide range of soils where the vegetation is, or was, deciduous forest. The soil is free-draining and has a pH of 5–7. It is rich in HUMUS and forms good agricultural land, particularly in southern and central England.

buffer state a neutral country between two powerful and usually warlike neighbours.

burn the Scottish word for a small stream.

bush veld a mix of open grassland and trees in varying proportions in Africa. It is a type of SAVANNA vegetation.

butte a steep-sided, flat-topped hill of layered strata (*see* STRATUM). The top layer is usually a resistant rock. It is found in dry or semi-arid areas. It is similar to, but smaller than, a MESA.

buttress a rugged, rocky ridge or face that projects from the side of a mountain.

Buys Ballot's Law in the northern hemisphere if a person has his or her back to the wind, the region of lower pressure will be on the left. This implies that, in the northern hemisphere, winds blow anticlockwise round a DEPRESSION and clockwise round an ANTICYCLONE. The converse is true in the southern hemisphere.

C

cadastre a record of an area including location, boundaries, value and ownership of the land. A *cadastral map* is thus a cartographic representation and is customarily drawn at a large scale.

cairn a Scottish Gaelic term for a pile of stones erected to mark a route, a boundary or to serve as a memorial. It also occurs in various Scottish place names, e.g. the Cairngorm Mountains or the village of Cairnbulg.

caldera a large depression at the centre of a VOLCANO with a diameter many times that of the original vent. It may be formed when a violent eruption destroys the top of the volcano or if the volcanic cone collapses inwards or if the top of an extinct or dormant volcano is destroyed by erosion. In time, vegetation grows in the caldera, and it may even become the site of a settlement, as in the Caldera de Freiras in Madeira.

Caledonian orogeny a period of mountain-building that occurred during the late Silurian/early Devonian periods, roughly 430–360 million years ago. It affected an enormous area, including Greenland, Scandinavia, Scotland and Ireland.

calf a piece of floating sea ice that has broken away from a larger piece of sea ice or from land ice. It also refers to a small island beside a larger one, as in the Calf of Man beside the Isle of Man.

California Current a cold OCEAN CURRENT flowing southwards along the west coast of the USA. It has a cooling effect on the CLIMATES of Oregon and northern California. Sea fogs form off San Francisco where it meets warmer conditions.

Campbell-Stokes recorder *see* SUNSHINE RECORDER.

canal (1) an artificial watercourse constructed for inland navigation or irrigation. (2) a long narrow arm of water connecting two larger seas. (3) an underground cave passage that is filled with water.

Canaries Current a cold current that flows southwards past the Canary Islands and the north coast of Africa. *See* OCEAN CURRENTS.

Cancer, Tropic of the imaginary line round the Earth at latitude 23° 32´N, where the Sun's rays are vertical at noon on 21 June.

canyon a steep-walled GORGE OR RAVINE that is much deeper than it is wide. It is cut by river action in areas of low rainfall where the river has a plentiful supply of water from a source higher up. The lack of rainfall means that there is little EROSION of the rocky sides of the canyon. The most famous example is probably the Grand Canyon in the USA.

Capricorn, Tropic of the imaginary line round the Earth at latitude 23° 32´S, where the Sun's rays are vertical at noon on 21 December.

cap rock an impervious rock overlying a source rock containing hydrocarbons. The cap rock contains the oil or gas and prevents upward migration. Typical cap rocks are limestone, shale evaporite and clay-rich sandstone.

carbon cycle the circulation of carbon compounds in the natural world by various metabolic processes of many organisms. The main steps of the carbon cycle are:

(1) carbon dioxide present in air and water is taken up during photosynthesis in plants and some bacteria;

(2) the carbon accumulated in plants is later released during the decomposition of the dead plant, or of bacteria or animals that have consumed any of the plant;

(3) carbon will also be released by the burning of fossilized plants in the form of fuels—coal, oil and gas—and during the respiration of all organisms. The concentration of carbon dioxide in the ATMOSPHERE is increasing as huge areas of tropical forests are destroyed while the consumption of fossil fuels is rising, i.e. less photosynthesis to absorb the increasing CO_2 level. This may be a factor involved in the small temperature rises throughout the world, known as the greenhouse effect. When there are high levels of carbon dioxide in the atmosphere, heat radiation from the Sun tends to be reflected back to Earth rather than lost to space.

cardinal points the major points of the compass: north, south, east and west.

carse a Scottish term for the fertile, low-lying alluvial ground around the river estuaries of eastern Scotland.

cartography the evaluation and compilation of all forms of data for, and the design and draughting of, a new or revised map. Cartography also includes the study of maps, methods of presentation and map use.

cartouche a highly decorative key on a map or CHART.

cataract a step-like series of waterfalls, cut by a RIVER as it crosses bands of hard rock. The water is fast-flowing and turbulent.

catchment area (1) the region drained by a river and its tributaries. (2) the area served by a city. (2) the area that receives a particular service, e.g. the catchment area of a school.

cave a large, natural underground chamber. The biggest caves are usually in limestone country and may contain spectacular stalactites (which grow down from the ceiling) and stalagmites (which grow up from the floor). Stalactites and stalagmites are formed when a solution of limestone evaporates slowly, leaving a deposit of calcium carbonate either on the cave roof or on the floor where a drop has fallen. RIVER-worn caves are usually bigger than sea-eroded caves.

cay *or* **key** a small, flat island made of sand overlying a CORAL REEF, just above high tide. Cays are particularly common in the Gulf of Mexico and on the coast of Florida.

Celsius scale or **centigrade scale** a temperature scale with a freezing point of 0° and a boiling point of 100°, devised by the Swedish astronomer Anders Celsius (1701–44).

centripetal drainage system a pattern of streams arranged radially and draining inwards towards a single RIVER or LAKE. The rivers that drain into Lough Neagh in Northern Ireland form a centripetal DRAINAGE system.

chalk a white, fine-grained, porous limestone formed from calcium carbonate and calcareous skeletal remains of micro-organisms. Chalk deposited in the Upper Cretaceous covers much of northwest Europe.

chalybeate spring water containing a high concentration of dissolved compounds of iron. It is reputed to be of therapeutic value and is therefore a feature of spa towns.

Chandler wobble the small wobble of the Earth on its axis during rotation. The amplitude of the wobble is about 0° 5′ and the period of about 14 months. The reason for the wobble is not understood but it is thought that it may be linked to EARTHQUAKES.

channel a water course that may be natural or manmade, including a RIVER channel that is shaped by the river itself, a CA-

NAL or a STRAIT of water, such as the English Channel, that links seas.

chaparral a poor grassland with short, dense bushes occurring in California, New Mexico and parts of South America between latitudes 30° and 40° S. Sometimes it contains trees such as evergreen oaks or pines. In all vegetation, the leaves are thick, hairy or leathery to prevent water loss. Chaparral develops where winters are mild and wet and summers are hot and dry. *See also* MAQUIS.

chart (1) a map of a coastline and its adjoining seas used for navigation. (2) a special map used for aviation. (3) a weather map to assist forecasting.

chasm *see* CANYON.

chelation a reaction between a metal ion and an organic molecule that produces a closed ring, thus tying up the unwanted metal ion. Chelation occurs naturally in soils, removing metal ions in solution that may be potentially toxic to plants. This principle can be applied to domestic products, e.g. chelating agents are often added to shampoos to soften water by 'locking up' calcium, iron and magnesium ions.

chimney (1) a vertical shaft leading up from a CAVE. (2) a vertical cleft in a rock wall. (3) in geology, a volcanic vent.

china clay a CLAY, composed primarily of kaolinite, that is formed by hydrothermal alteration of granite. It is extracted using high pressure water jets and is used in many industries, including ceramics, paper and pharmaceuticals.

chinook a warm, dry wind blowing down the eastern side of the Rocky Mountains, usually between December and February. It occurs when moist air blows in from the Pacific to replace the normal HIGH PRESSURE over the Prairies. The chinook is adiabatically (*see* ADIABATIC TEMPERATURE CHANGE) warmed as it descends from the mountains and can raise the temperature dramatically, causing AVALANCHES and rapid thaws. *See also* BERG WIND; FÖHN.

cinder cone a cone surrounding a volcanic vent composed of small fragments of glassy, solidified lava.

circuit of capital the capitalist invests money to pay for the production of goods. Part of the production process involves paying wages to the workforce. Wages enable workers to buy goods. The profit on sales provides further capital for investment.

circumference of Earth the equatorial circumference is 40,076 kilometres (24,902 miles). The circumference around the poles is 40,008 kilometres, (24,860 miles).

cirque a circular hollow at the head of a VALLEY or GLACIER formed by the EROSION effect of a glacier. The sides and back are steep but the front opens out downslope to the valley. The cirque may be dry or it may contain water (a *cirque lake*) or a *cirque glacier*. Cirques are usually 1 kilometre or less across and the depth is usually about one-third the extent.

cirrocumulus a type of high-level cloud formed in stable air as sheets or layers with ripples and waves.

cirrostratus a type of cloud composed of whitish or near-transparent veils with a smooth or fibrous form. Often this cloud type covers the sky, producing a rainbow or white ring around the Sun or Moon due to refraction by ice crystals.

cirrus a high-level cloud forming narrow bands or filaments.

clay a fine-grained sediment composed of clay minerals that are hydrous aluminium silicates. Clay is used extensively in numerous industries, including the manufacture of ceramics and bricks; rubber, plastic and paint production; as fillers in paper manufacture, and in drilling muds.

clear felling the cutting down of all trees on a site. It is not a good practice as it leaves the soil unprotected against EROSION.

climate characteristic weather conditions produced by a combination of factors, such as rainfall and temperature. Whether taken singly or jointly, these factors, together with influencing features such as altitude and latitude, produce a distinctive arrangement of zones around the Earth, each with a generally consistent climate when studied over a period of time. The major climatic zones are, from the Equator:

humid tropical—hot and wet

subtropical, arid and semi-arid—desert conditions, extremes of daily temperature

humid temperate—warm and moist with mild winters

boreal (northern hemisphere)—long cold snowy winters and short summers

subarctic (or *subantarctic*)—generally cold with low precipitation

polar—always cold

Climatic patterns are very dependent on heat received from the Sun, and at the poles the rays have had to travel farther and, in so doing, have lost much of their heat, producing the coldest regions on the Earth. Conversely, air near the Equator is very warm and can therefore hold a great deal of water vapour, resulting in hot and wet conditions—humid tropical. There are more complex systems of climate classification, e.g. KÖPPEN and THORNTHWAITE, which are based on PRECIPITATION and EVAPORATION, characteristic vegetation and temperature. In each system, the Earth can be split into numerous provinces and smaller areas, producing quite a detailed overall picture.

climatology the scientific study of the climates of the Earth, including their origin, variation and regional and global effects of climate on the environment and life on Earth.

cloud clouds are droplets of water or ice formed by the condensation of moisture in a mass of rising air. Water vapour formed by evaporation from seas, lakes and rivers is ordinarily contained in air and becomes visible only when it condenses to form water droplets. Warm air can hold more water vapour than cold air, thus when air rises, becoming cooler, it becomes saturated ('full') of water and eventually droplets form, as cloud, each droplet forming around a central nucleus such as dust, pollen or a smoke particle. Clouds are classified firstly upon their shape and then by their height. There are three major groups: *cumulus* ('heap' clouds), *stratus* (sheet-like) and *cirrus* (resembling fibres), which are divided further, as cloud forms show a mix of shape, e.g. cirrocumulus.

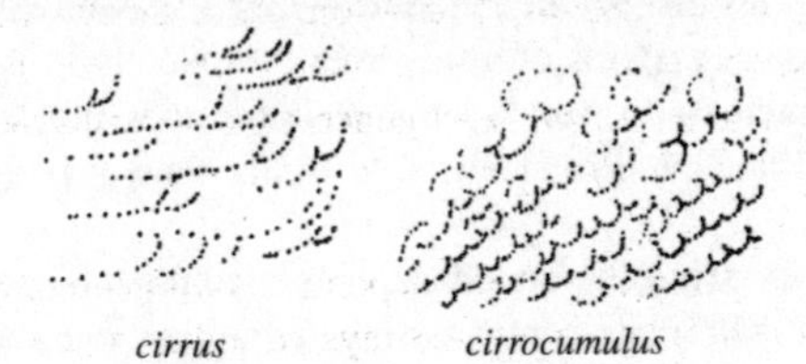

cirrus *cirrocumulus*

stratus *cumulus*

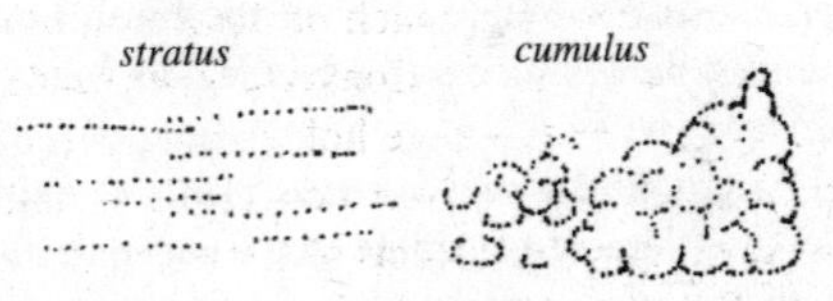

coire, corrie the Gaelic and Scottish terms for a CIRQUE.

col (1) the lowest point, leading from one valley to another, between the peaks on a mountain RIDGE. (2) an area of low pressure and low wind speed connecting two areas of lower pressure and lying between two diametrically opposed ANTICYCLONES.

cold desert (1) polar regions where plant life is hindered by low temperatures and lack of free moisture. (2) the enclosed basins of central Asia where mean temperature is less than 6°C (42°F) for at least one month and there is little rainfall.

cold front the line defining the front of a cold air mass that will cut under warm air ahead of it. The weather changes that may occur include a fall in the temperature and veering often squally wind.

cold water desert an arid but often foggy desert where the CLIMATE is affected by a cold OCEAN CURRENT, e.g. western California. Incoming air streams cool as they pass over the cold current and rain falls out at sea, leaving a dry wind to blow inland.

collective farm a large, state-owned farm made by merging small farms that belonged to individuals. They are a feature of Communist systems of government. Productivity is poor because increased effort by the workers does not lead to increased income. Produce is sold to the state at fixed rates of pay.

collision margin the boundary of two continental plates. It is thought that, if two continental plates collide, one slides under the other, causing a double layer of the Earth's crust. In this way mountain ranges such as the Himalayas were formed (*see also* PLATE TECTONICS).

colony (1) a group of closely related plants or animals in one place. (2) a settlement of people in a foreign country who are ruled by their home country.

combe *or* **coombe** (1) a dry valley in the chalky areas of southern England that ends in a steep-sided hollow. In Devon, it is a short valley with a stream that descends steeply to the sea. (2) in the Lake District, a CIRQUE. The word is widely used in English place names.

common land land that is owned privately by an individual or organization but the legal rights of which are held by others.

common market an association of states into a single trading market that has little or no restriction on trading between the states and has a united policy on trading with non-member states. *See* EC; FREE MARKET.

compass an instrument that indicates direction. A magnetic needle swings freely in a horizontal plain, thereby seeking the magnetic north or south pole by following local MAGNETIC DECLINATION. The needle moves across a graduated dial that shows the CARDINAL POINTS. Thus direction of north can be established. In practice, care must be taken to avoid local magnetic fields, e.g. ships' compasses are adjusted to compensate for interference from any magnetic fields that may be set up by equipment nearby. *See* MAGNETIC DEVIATION.

compiled map a map produced by using information from an existing map and not from an original survey.

composite map usually a COMPILED MAP that uses information from several other maps and displayed to enable comparisons to be made between sets of data.

concordant coast a coast that lies parallel to the TOPOGRAPHY of the area and itself is determined by the underlying geology. It may be a linear coastline or a series of islands and promontories lying parallel to each other, e.g. the Dalmatian coast of the former Yugoslavia. It is also known as a *Pacific type coast.*

condensation (1) the process by which a substance changes from the gaseous state to the liquid state and, in so doing, loses KINETIC ENERGY. (2) in meteorology, the process of forming a liquid from its vapour. If moist air is cooled below its DEW POINT and nuclei or surfaces are available, then water vapour will condense. The nuclei may be particles of dust or ions. In meteorology condensation may also be caused by a change in volume and temperature of the air; meeting a colder body whether air or a solid; and the expansion of air without further heat input.

conglomerate (1) a coarse-grained rock containing clasts (rock fragments) that are rounded or sub-rounded and larger than 2 millimetres (.079 of an inch). (2) a firm controlling a group of industries producing unrelated products. For example, Hanson, in 1996, owned brick-making companies and electricity companies, amongst others. There is presently a tendency for conglomerates to sell off companies and to become smaller, but such trends are often changed or reversed.

coniferous forest a forest of needle-leaved trees that are usually evergreen and shallow-rooted and bear cones. Conifers are fast-growing and are therefore planted as a source of softwood timber and pulp. They are tolerant of a wide range of soils, CLIMATES and terrain and are therefore widespread, from sub-polar to tropical latitudes and from mountainous to coastal areas.

conjunction two heavenly bodies lying in a straight line relative to the Earth. When the Sun and the Moon are in conjunction on the same side of the Earth, maximum force is exerted on the tide, and SPRING TIDES occur.

conservation the protection of species and of natural or man-made resources and landscape for present and future use. It should not be confused with preservation. The conservationist supports managed exploitation of a stock at a sustainable level. The preservationist tends to ban the use of a stock altogether.

contiguous zone a zone beyond the territorial seas over which a nation claims exclusive rights. In law, the contiguous zone stretches from 12 to 24 nautical miles beyond the coastline.

continent one of several large landmasses covering 29 per cent of the Earth's surface, of which 65 per cent is in the northern hemisphere. Where the continent edge meets the sea is a CONTINENTAL SHELF and slope that is often cut by submarine canyons. Farther oceanward lies the deepest parts of the ocean, the ABYSSAL plain. There are seven continents making up the Earth.

	highest point		*area*	
	metres	*feet*	*square kilometres*	*square miles*
Asia	8848	29,028	43,608,000	16,833,000
Africa	5895	19,340	30,335,000	11,710,000
North and Central America	6194	20,320	25,349,000	9,785,000
South America	6960	22,834	17,611,000	6,798,000
Antarctica	5140	16,863	14,000,000	5,400,000
Europe	5642	18,510	10,498,000	4,052,000
Oceania	4205	13,796	8,900,000	3,400,000

continental air mass a usually dry AIR MASS originating over a continental interior where high atmospheric pressure occurs. It may form in low or high latitudes.

continental climate the type of CLIMATE found in the interiors of large land masses in mid-latitudes. Rainfall is low, with a summer maximum when CONVECTION RAIN falls. Winters are extremely cold and summers very hot.

continental crust the Earth's crust, which lies beneath the CONTINENTS and CONTINENTAL SHELVES. It is 30–40 kilometres (19–25 miles) thick mostly, but increases to approximately 70 kilometres (44 miles) beneath areas of mountain-building. The crust has two layers of differing compositions and slightly different densities, and its base is marked by the Mohorovicic Discontinuity, where both composition and density change significantly.

continental drift a geological concept, formulated by the German geophysicist Alfred Wegener (1880–1930), that 200 million years ago the Earth consisted of a large single continent, called Pangaea, that broke apart to form the present continents. An explanation of how such huge land masses move is provided by studying the vast plates that make up the outer layer of the Earth, called the crust. The crustal plates are believed to float on a partially molten region of the Earth between the crust and the Earth's core, the lower mantle (hence the modern term, PLATE TECTONICS).

continental shelf the surface between the shoreline and the top of the CONTINENTAL SLOPE, where the gradient steepens at a depth of approximately 150 metres (*c*.490 feet). The average width of the shelf is 70 metres (*c*.230 feet).

continental slope the slope between the edge of the CONTINENTAL SHELF and the deep sea floor. It is very steep with a gradient of 2°–5°. There is considerable sliding and slumping of marine sediments on the slope. It makes up about 8.5 per cent of the ocean floor.

contour a line connecting points on the surface that are the same height above datum (reference surface).

contour ploughing a method of ploughing parallel to the contours instead of up and down a slope. Contour ploughing helps to prevent EROSION and the formation of GULLIES.

control point *or* **control station** a point on the ground whose exact position and elevation has been determined with great accuracy. Control points or stations are used as reference points from which other points can be plotted accurately when carrying out survey work.

conurbation a group of large towns or cities that have expanded towards each other so that they form a continuous built-up area.

convection a method of heat transfer through liquids and gases. It will occur only if the lower temperature area is above the high temperature area of a liquid or gas. CONVECTION CURRENTS can be seen easily when a dye is placed at the bottom of a container of water and heat is applied to that region, causing the dye and water molecules to rise and disperse throughout the container. An everyday example of convection that is readily detected by hand is the hot air that rises from a warm radiator.

convection current any current of warm fluid, gas or liquid that rises and is replaced by colder fluid. The warm fluid rises because it is less dense than the colder fluid. In geography, the fluids under consideration are air and water. Convection currents play a big part in influencing CLIMATES and in the circulation of OCEAN CURRENTS.

convection rain when moist air is warmed and rises as a CONVECTION CURRENT, it cools adiabatically (*see* ADIABATIC TEMPERATURE CHANGE). If its temperature falls below DEW POINT, clouds, often CUMULONIMBUS, form and heavy rain falls. Convection rain is associated with EQUATORIAL CLIMATES and the cold FRONTS of POLAR AIR MASSES. It may also occur in summer in continental interiors.

coombe *see* COMBE.

coral reef a hard bank built up by the carbonate skeletons of colonial corals (and algae). There are several forms of reef, including *barrier reefs*, *fringing reefs* (which are attached to the coast) and ATOLLS (where a reef encloses a LAGOON). Certain conditions are essential for the growth of reefs, including temperature, a maximum water depth of 10 metres (33 feet), clear water with no land-derived sediment, and normal marine salinity.

co-range line a line drawn on a map joining either climatic stations showing the same temperature range between January and July or points with an equal tidal range.

cordillera a Spanish term meaning a major system of MOUNTAINS along with the VALLEYS and PLATEAUX of the area, e.g. the Rockies. Cordilleras are folded mountains, formed at destructive plate boundaries.

core sampling a method of withdrawing samples of soil, peat, rock or ice from the area under examination. It is the main method of determining the mineral wealth of rock strata or of determining the load-bearing capacity of rock in civil engineering.

Coriolis force when dealing with moving objects in relation to rotating systems (e.g. a particle moving away from an observer on the Earth), although the particle moves in a straight line, it will appear to the observer to move in a curved path. The Coriolis force is a theoretical force that is used to simplify such calculations, e.g. in calculating movement of air on the Earth's surface.

corrie *see* COIRE.

costa the Spanish for coastline.

côte the French for (1) a coast. (2) a steep slope.

coulée (1) a congealed, glassy LAVA flow with steep sides. (2) a GORGE gouged out by the sudden, violent release of water from a dammed lake. (3) a tongue of fine debris formed by PERIGLACIAL processes.

couloir a narrow GULLY with a steep slope in precipitous mountain terrain.

cove a small BAY in a rocky coast.

crag a steep rock face or BUTTRESS on a mountainside.

crater (1) a circular depression with high, steep, inward facing cliffs round the vent of a VOLCANO. It may contain a *crater lake*. (2) the circular depression caused by the impact of a meteorite.

creek (1) a tidal channel in a coastal marsh or between sandbanks. (2) a small stream.

creep the slow gravitational movement of soil, SCREE or glacial ice down a slope.

crevasse a deep vertical or wedge-shaped fissure in a GLACIER.

It is caused by stresses within the moving ice as, for example, when moving over a small hill. It can vary in width from a few centimetres to tens of metres. It allows meltwater and rock DEBRIS to penetrate into the glacier. There are several types of crevasse—radial, marginal, etc—depending upon their position with respect to the glacier.

crevice (1) a narrow crack in rocks. (2) a mineral-bearing vein of rock.

cruciform village a village that has formed at the intersection of two different routes.

crude oil *see* PETROLEUM.

crust of the Earth the outer layer of the Earth's structure, which is between 6 and 48 kilometres thick. It includes the continents and ocean floor, and is essentially the LITHOSPHERE.

cuesta an asymmetrical RIDGE with a long, gentle slope on one side and a short, steep slope or ESCARPMENT on the other. It is produced when the slopes erode at different rates. The gentle slope follows the DIP of the underlying STRATA, which are resistant to EROSION.

cumulonimbus large, bulging clouds that reach great heights, with the upper reaches forming anvil shapes or plumes. The base is dark, usually producing precipitation.

cumulus well-defined CLOUD in the form of separate masses comprising bulges and towers with a darker, often flat, base.

current the movement of a fluid in a particular direction. The rate of flow of river currents vary, not only with the steepness of the ground over which they flow but also with the width and depth of the river, due to friction along the banks and the river bed. The velocities of ebb and flow tides vary with the nature of the coastline. Rip tides are fast-flowing near shore currents. OCEAN currents have an important effect on CLIMATE, as do air currents. The wind is simply an air current.

cwm the Welsh term for CIRQUE.

cyclogenesis the formation of CYCLONES. This occurs in specific areas, such as the western North Atlantic, the western North Pacific and the Mediterranean Sea. When air in the upper atmosphere disperses more quickly than it can be replaced, an area of very low pressure results, forming a cyclone, and a period of stormy weather follows while wind from a high pressure area flows in to fill the DEPRESSION.

cyclone a low pressure area with strong WINDS. In many areas the term is being replaced by DEPRESSION.

cyclothem a type of cyclical sedimentation that is created by relatively rapidly changing conditions of deposition. There is a continuing change from freshwater to brackish water and then marine conditions, and back to brackish and finally fresh water. In each situtation, differing sediments are produced. The term is also applied to the sequence fresh-brackish, marine-fresh-brackish-marine, etc.

cymatogeny warping of the Earth's crust on a massive scale, resulting in domes and basins. The domes can collapse with faulting to form RIFT VALLEYS.

D

dale a valley, often containing a stream or RIVER and common in many Scottish and north of England place names.

Dalmatian-type coast an example of Pacific or CONCORDANT COAST.

dambo a shallow depression at the head of a DRAINAGE system but one that has no identifiable drainage streams.

datum level *or* **datum line** the zero level or line from which all land altitudes and water depths are calculated. It is usually taken as sea level or a mean based on levels of the tide. In Britain it is the ORDNANCE DATUM.

dead caves caves that were made by the sea but are now above sea level because the sea level has fallen or the land has been raised. The caves are therefore no longer subject to marine EROSION.

dead cliffs cliffs that were eroded by the sea but are now above sea level (*see* DEAD CAVES) and are no longer subject to marine EROSION.

debris material such as SCREE, gravel, sand or CLAY, formed by the breaking up of rocks, that has been moved by ice or water from its original site to another location.

declination *see* MAGNETIC DECLINATION.

declining region a region suffering the economic problems associated with factory closure, unemployment, old-fashioned working methods etc. *See* MULTIPLIER.

deep a long narrow trench over 5000 metres deep in the ocean floor. They tend to be found close to ISLAND ARCS or along coasts bordered by high mountain ranges, where they are called *foredeep*. They are thought to be the result of one TECTONIC PLATE being pushed under another where they met, i.e. they occur at destructive plate margins.

defile a narrow pass, similar to a GORGE in a mountainous area.

deflation hollow a large basin formed by deflation, i.e. the action of the wind in removing loose CLAY, SILT or sand from the surface of the ground. Deflation hollows usually occur in dry areas. If enough material is removed so that the WATER TABLE is reached, the hollow may form an OASIS.

delta a roughly triangular area of sediment formed at the mouth of a RIVER. It is caused by a current laden with sediment entering a body of water, resulting in a reduction of the current's velocity and thus its carrying capacity. Much of the sediment is therefore deposited on entering the LAKE or sea. The shape of the delta depends on various factors, including water discharge, climate, tides and sediment loads. Modern examples include the Mississippi and the Nile.

demand the volume of goods and services that people are willing to buy. In general, as prices rise, demand falls, and vice versa.

demography the statistical study of human population. It is concerned with the size, distribution and composition of the population at various age ranges.

depression an area of low atmospheric pressure with closely packed ISOBARS producing a steep pressure gradient and therefore very strong winds. Because of the Earth's rotation, the winds circulate clockwise in the southern hemisphere and anticlockwise in the northern hemisphere. In tropical regions, cyclones (tropical HURRICANES) combine very high rainfall with destructive winds, which results in widespread damage and possible loss of life.

Outside the tropics, in more temperate climates, the term cyclone is being replaced by depression (or *low*) in which the pressure may fall to 940 or 950 mb (millibars). Compare this with the average pressure at sea level, which is 1013 mb. *See also* ANTICYCLONE.

desert an ARID or semi-arid (i.e. dry and parched with under 25 centimetres/10 inches of rainfall annually) region in which there is little or no vegetation. The term was always applied to hot tropical and subtropical deserts, but is equally applicable to areas within CONTINENTS where there is low rainfall and perennial ice-cold deserts.

The vegetation is controlled by the rainfall and varies from sparse, DROUGHT-resistant shrubs and cacti to sudden blooms of annual plants in response to a short period of torrential rain. If the groundwater conditions permit, e.g. if the WATER TABLE is near to the surfac,e creating a spring, or the geology is such that an artesian well (*see* AQUIFER) is created, then an OASIS may develop within a hot desert, providing an island of green.

Hot deserts are found in Africa, Australia, United States, Chile and cold deserts in the Arctic, eastern Argentina and mountainous regions.

Hot desert extremes

extreme of shade temperatures: Death Valley, California, maximum 58°C; maximum daily range 41°C
maximum ground surface temperature: Sahara, 78°C
extreme of rainfall: Chicama, Peru, 4 mm per year

The process whereby desert conditions and processes extend to new areas adjacent to existing deserts is called *desertification*. *See also* COLD DESERT; HOT DESERT.

desert pavement a wide area of wind-polished, rounded stones in a DESERT. It is formed when the sand has been blown away but it protects the underlying ground from further EROSION.

desiccation a progressive increase in aridity of an area. It may be the result of natural causes, such as a change of climate. More often it is the result of human activity, such as deforestation, overstocking with grazing animals, changing the course of a RIVER, etc. During desiccation, the WATER TABLE is likely to fall, and the area will become a DESERT.

detrital mineral a mineral derived from a parent ROCK by the mechanical breakdown of the rock by WEATHERING and EROSION. Diamond, zircon and gold are typically resistant minerals.

development the use of resources to improve the standard of living. The government may provide grants and other incentives to encourage development in a rundown area, e.g. some rundown inner cities at present enjoy development status. Government may provide similar incentives to encourage development in remote areas.

dew condensation of water vapour in the air, producing water or moisture, due to the temperature falling causing the vapour to reach saturation. Surfaces become cooled by radiation, and when the temperature goes below the DEW POINT, condensation occurs from the air in contact with the cool surfaces.

dew point the temperature at which air becomes saturated with water vapour and deposits DEW.

dike *see* DYKE.

dip the angle of inclination of a plane, measured from the horizontal and perpendicular to the strike. It applies particularly to rock outcrops where the orientation of the bedding or cleavage is measured.

discordant coast *see* ATLANTIC-TYPE COAST.

discordant drainage *see* ACCORDANT DRAINAGE.

discordant junction a junction where a TRIBUTARY falls steeply just before it joins a RIVER.

distributary a branch of a RIVER that leaves the main river and does not return to it. Distributaries form in DELTAS where the ground is relatively flat and the main stream forms many branches among the silty deposits as the water flows towards the sea.

diurnal range a measure of the difference between maximum and minimum temperatures recorded in one 24-hour period. It is greatest in continental DESERT areas, far from the moderating influence of the sea, where days are very hot and nights very cold.

diversification the spread of industry over a wide range of activities so that there is no overdependence on one.

doldrums relatively calm OCEAN areas around the EQUATOR, which have low pressure and light winds.

donga (1) in South Africa, a steep-sided GULLY produced by soil EROSION, often the result of floods. (2) in Australia, a circular depression left when the roof of a CAVE has collapsed.

dormitory town a town where the majority of the residents work elsewhere. Dormitory towns usually grow round the edges of established villages. As people tend to shop in the town in which they work, however, retail outlets and other services tend to be lacking.

downs gently undulating landscape formed over CHALK hills in the south of England. Downs may be grassland or used for arable farming.

drainage the movement of water, derived from rain, snowfall and melting of ice and snow, over the land (and through it in subterranean waterways) that results in its eventual discharge into the sea. The flow of streams and RIVERS is influenced by the underlying rocks, how they are arranged and whether there are any structural features that the water may follow. Further factors affecting drainage include soil type, climate and the influence of humankind. There are a number of recognizable patterns that can be related to the geology, and these are illustrated on page 50. When a drainage pattern is a direct result of the underlying geology, it is said to be *accordant* (the opposite case being *discordant*).

drought a long, continuous period with little or no PRECIPITATION. In Britain, a drought is officially defined as a period of at least 15 consecutive days on none of which there is more than 0.25 millimetre (0.01 inch) of rain.

drove road a broad track, used by herders to walk their animals to markets in towns or cities before other means of transport were available.

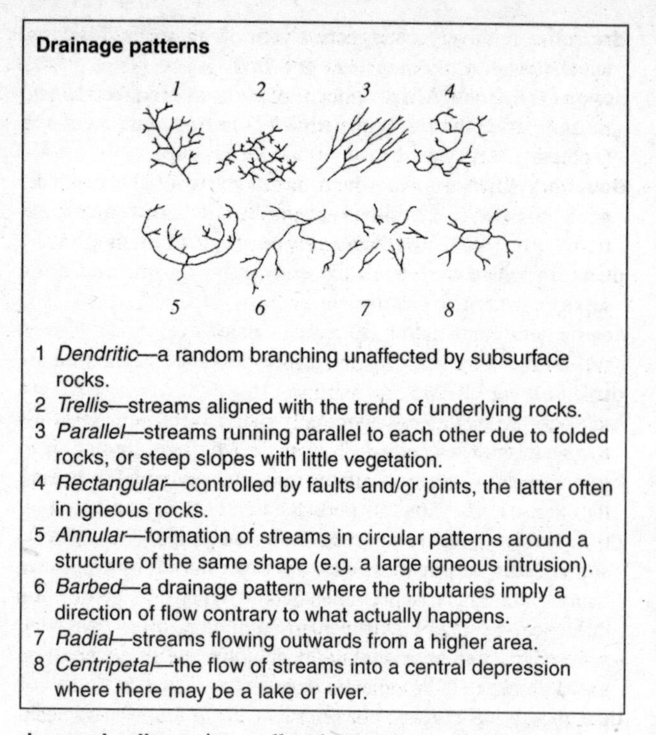

Drainage patterns

1 *Dendritic*—a random branching unaffected by subsurface rocks.
2 *Trellis*—streams aligned with the trend of underlying rocks.
3 *Parallel*—streams running parallel to each other due to folded rocks, or steep slopes with little vegetation.
4 *Rectangular*—controlled by faults and/or joints, the latter often in igneous rocks.
5 *Annular*—formation of streams in circular patterns around a structure of the same shape (e.g. a large igneous intrusion).
6 *Barbed*—a drainage pattern where the tributaries imply a direction of flow contrary to what actually happens.
7 *Radial*—streams flowing outwards from a higher area.
8 *Centripetal*—the flow of streams into a central depression where there may be a lake or river.

drowned valley a river valley that has been submerged as a result of a rise in sea level, often caused by post-glacial melting.

drumlin an oval accumulation of boulder CLAY that forms a small hill. Typically, it has one end blunt and the other pointed because of streamlining by the movement of the ice from which they were deposited. They are commonly found in groups, with their line of elongation reflecting the movement direction of the ice.

dry valley as the name suggests, a valley feature originally created by water erosion but now dry. This may be the result of a fall in the WATER TABLE, removal of earlier permafrost conditions, or to 'capture' of the flow by another river or stream (*see* RIVER CAPTURE).

dual economy a system found in many developing countries where a relatively advanced economy exists alongside a traditional economy but the two rarely come into contact.

dune an accumulation and movement of sediment, usually sand, by the wind. The size varies, but in the Sahara Desert can rise to 300 metres (*c*.1000 feet) in height. There are several types of dune, depending on morphology and formation.

dust bowl in the semi-arid southwest of the USA lack of WINDBREAKS, overgrazing and generally poor land management led to soil EROSION, when the protection of the grass was lost by ploughing. The dry topsoils were blown away, leaving areas that became known as dust bowls.

dust devil a small, short-lived WHIRLWIND that picks up loose soil or sand and sweeps it round a central vortex. They are caused when land in dry areas gets very hot and CONVECTION CURRENTS form. The swiftly rising air currents form a whirlwind that, although it does not do much damage, can travel at speeds of up to 32 kilometres an hour (20 mph).

dust storm a storm in dry or semi-ARID areas where great quantities of dust are whipped up by the wind, causing serious loss of visibility. Dust can be carried to great heights, temperatures are very high and there is high electrical tension. Dust storms may advance either as a 'wall' or as a WHIRLWIND. Both are a serious danger to air traffic.

dyke *or* **dike** (1) igneous intrusion that occurs as a sheet-like body with near-parallel sides. Dykes are normally discordant, cutting across the country (host) rock, and are usually vertical or nearly so. (2) an artificial embankment to prevent flooding. (3) an artificial drainage ditch.

dynamic equilibrium a state of balance despite the physical changes taking place. For example, a SPIT may appear to be unchanging while it is losing material from its seaward side but gaining material on its landward side.

E

Earth the third planet in the solar system with its orbit between Venus and Mars. It is a sphere, flattened at the poles, with an equatorial radius of 6378 kilometres (3963 miles). It comprises a gaseous ATMOSPHERE, a liquid HYDROSPHERE and solid LITHOSPHERE. The core is part liquid and has a relative density of 13, and a temperature of almost 6500K. It is composed of nickel-iron separated into an inner and outer core.

earthquake the often violent movement of the Earth as a result of tectonic upheaval caused by the sudden release of accumulated stress, perhaps along a fault or fault zone. Earthquakes are classified by depth of focus: shallow—less than 70 kilometres (43.5 miles); intermediate—70–300 kilometres (43.5–186.4 miles); deep—more than 300 kilometres (*see* EPICENTRE; RICHTER SCALE). Attempts to predict earthquakes rely on measurement of stress increases but can so far be related only to an increasing risk of activity.

Earth's magnetic field *see* GEOMAGNETISM.

earth tremor a small earthquake that is unlikely to cause any damage.

easting the first part of a GRID REFERENCE, always preceding a NORTHING, when map coordinates are quoted. It represents dis-

areas of earthquake activity

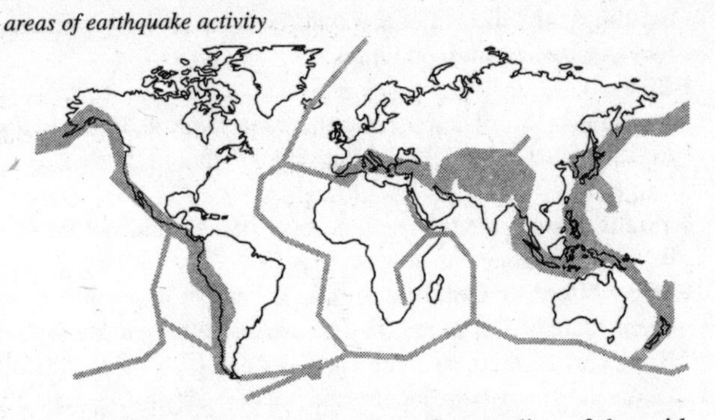

tance measured eastwards from the reference line of the grid. *See* NATIONAL GRID.

ebb tide the outflowing stream of sea water after high tide. In some areas the ebb tide is a powerful current.

EC *or* **EEC** abbreviation of European Community or European Economic Community (formerly Common Market). It is a trading area whose members are Britain, Ireland, Austria, Belgium, Denmark, Finland, France, Germany, Greece, Italy, Luxembourg, the Netherlands, Portugal, Spain and Sweden.

ecology the study of the relationship between plants and animals and their environment. Ecology is also known as *bionomics* and is concerned with, for example, predator-prey relationships, population dynamics and competition between species.

ecosystem an ecological community that includes all organisms that occur naturally within a specific area.

eddy a roughly circular movement within a body of air or water. Eddies can be much stronger than the smooth CURRENT in which they form and arise as the current moves over or round some irregularity in its path. Eddies form on the lee side of buildings, and their effects should be taken into consideration when designing high buildings.

EEC *see* EC.

effluent the flow of sewage, fertilizers in solution, liquid industrial waste or other liquid pollution into streams, RIVERS, LAKES or the sea.

effluent stream a small DISTRIBUTARY that should not be confused with effluent as in POLLUTION.

Ekman effect *or* **Ekman spiral** as the wind blows across the ocean there is a tendency for the surface water to move with it. However, the rotation of the Earth deflects the flow to the right in the northern hemisphere and to the left in the southern hemisphere. As depth of water increases, the effect of the wind and surface flow becomes less until at about 100 metres (328 feet) it ceases to have any effect and flow is influenced wholly by the Earth's rotation. This is the Ekman spiral.

The overall effect is that the body of water moves at about right angles to the prevailing wind. Where the prevailing wind blows parallel to the coast, the body of water is pushed away from the coast, allowing cold currents to well up. The nutrients in these are carried across the ocean in the warmer surface water and provide feed for fish shoals over a wide area.

electromagnetic waves the effects of oscillating electric and magnetic fields that are capable of travelling across space, i.e. they do not require a medium through which to be transmitted. The spectrum of electromagnetic waves is divided into the following categories:

	wavelength (metres)	*frequency (Hz)*
gamma rays	10^{-10}–10^{-12}	10^{21}–...
X-rays	10^{-12}–10^{-9}	10^{21}–10^{17}
ultraviolet radiation	10^{-10}–10^{-7}	10^{18}–10^{15}
visible light	10^{-7}–10^{-6}	10^{15}–10^{14}

	wavelength (metres)	*frequency (Hz)*
infrared radiation	10^{-6}–10^{-2}	10^{14}–10^{11}
microwaves	10^{-3}–10	10^{11}–10^{7}
radiowaves	10–10^{6}	10^{7}–10^{2}

All electromagnetic waves travel through free space at a speed of approximately 3×10^8 ms^{-1}, known as the *speed of light*. The only electromagnetic waves that are readily detected by the eye are visible light waves. These consist of various wavelengths that correspond to the colours red, orange, yellow, green, blue, indigo and violet. The colour red has the longest wavelength, lowest frequency, and the colour violet has the shortest wavelength, highest frequency. An object that appears to be red in colour, for example, has absorbed all the light waves from the blue end of the spectrum while reflecting the ones from the red end.

El Niño a south-flowing current along the west coast of South America from Colombia to Ecuador. Between Christmas (*El Niño* means Christ Child) and Easter, it flows south to Peru, so blocking the north-flowing, cold, nutrient-rich HUMBOLDT CURRENT. In some years, El Niño fails to retreat at Easter, the Humboldt Current continues to be blocked and cannot reach the fishing grounds, with disastrous results for the Peruvian fishing industry.

The reasons for the abnormal behaviour of El Niño are not understood, but it seems to be connected to a low pressure area that, instead of remaining over Indonesia, drifts out to the mid-Pacific Ocean. This causes westerly winds that pile up warm surface water against the coast of South America, thus enhancing the effects of El Niño. As the low pressure area drifts eastwards, it brings rain to Ecuador and Peru, while the high pressure area over Indonesia causes DROUGHT in Australia. *See* OCEAN CURRENTS.

emergent shoreline any shoreline that is now either raised above high water mark or is inland from the present coastline. They may be the result of uplift of the land or a fall in sea level.

enclave a small area within a country that is administered by another country.

enterprise zone an area in economic decline that is chosen for DEVELOPMENT. Private enterprise may be attracted by tax concessions, provision of buildings, relaxed planning regulations, etc.

entrepôt a point of entry where goods are imported without attracting duty before being sent on to another country. For example, Rotterdam is an entrepôt for much of Europe.

environment the total surroundings in which an organism lives. It includes land, water and air as well as more aesthetic concepts such as landscape, whether natural or artificial.

epicentre the point or line on the Earth's surface that is directly above the focus of an EARTHQUAKE.

equable climate a CLIMATE with little variation throughout the year, e.g. a MARITIME CLIMATE.

Equator the imaginary GREAT CIRCLE round the Earth at latitude 0°. It is 40,076 kilometres (24,902 miles) long.

equatorial climate the type of climate occurring on low ground near the EQUATOR. Temperatures and humidity are perpetually high, and there is little seasonal change except for rainfall. Day and night are of more or less equal length.

equatorial currents surface currents in the oceans near the EQUATOR that flow at about 2 knots. The north and south equatorial currents flow westwards, while between them the equatorial counter current flows eastwards. *See* OCEAN CURRENTS.

equatorial trough *see* DOLDRUMS.

equinox the point of intersection between the Sun's *apparent* path in the sky (relative to the stars) and the celestial EQUATOR. As the Sun physically crosses the celestial equator north to south, it is the *autumnal equinox*, and south to north is the *vernal equinox.*

erosion specifically, erosion involves the further breakdown and transport of material by water, ice and WIND, and in the transportation the continued wearing down of land surfaces. The transporting agents thus erode by wind laden with sand scouring ROCK, GLACIERS containing rocks and boulders grinding down the rocks over which they pass, and rivers excavating their own courses due to movement of rocks, pebbles and particles in the water. There are six different kinds of erosion processes;

abrasion	wearing away through grinding, rubbing and polishing.
attrition	the reduction in size of particles by friction and impact.
cavitation	characteristic of high energy river waters (e.g. waterfalls, cataracts) where air bubbles collapse, sending out shock waves that impact on the walls of the river bed (a very localized occurrence).
corrasion	the use of boulders, pebbles, sand, etc, carried by a river, to wear away the floor and sides of the river bed.
corrosion	all erosion achieved through solution and chemical reaction with materials encountered in the water.
deflation	the removal of loose sand and silt by the wind.

Rivers also carry materials in solution.

escarpment *or* **scarp** a cliff or steep slope at the edge of an essentially flat or gently sloping area, generated by a combination of original geology and the attitude of the rocks, and subsequent EROSION.

esker a steep-sided ridge composed of sands and gravels, the former showing cross-bedding. It is the remains of a stream that ran beneath or within a GLACIER. Eskers do not show any relationship to the modern pattern of DRAINAGE, and good examples can be found in Scotland and Scandinavia.

estuary a partially enclosed stretch of water that is subjected to marine tides and fresh water draining from land. An estuary is usually created as a DROWNED VALLEY, as a result of a post-glacial rise in sea level. A large amount of sediment is deposited in estuaries, and the tidal currents may produce channels, sandbanks and sand waves.

étang a small brackish LAKE among coastal sand DUNES or beach ridges.

ethnic group a group of people with their own distinctive culture and customs, living within a larger, different society and subject to the laws of that larger society.

euphotic zone the upper layers of the sea or LAKES where light can penetrate and, therefore, where life is possible since photosynthesis can take place.

eustasy a worldwide change of sea level that may be the result of, firstly, climatic change causing either the growth or melting of ice sheets or, secondly, a change in the shape of OCEAN BASINS.

eutrophication the enrichment of a LAKE or RIVER by receiving nutrients, usually effluent from silage, sewage or dissolved fertilizer. When large amounts of nutrients are added over a short time, a thick algal bloom grows and uses up all the oxygen from the water. Other forms of life then suffocate.

evaporation the process whereby a liquid is changed into a gas. Evaporation from the Earth is faster when the air is dry and warm, where there are strong winds and little vegetation cover.

everglade marshy ground with tall, coarse grass and occasional trees. Flooding usually occurs during the summer rainy season. The Florida Everglades are probably the best-known example.

exfoliation the process whereby rocks are gradually worn away by the flaking off of layers or shells. The process involves a number of factors, including chemical WEATHERING, which breaks down and may cause expansion of some minerals, and repeated expansion or contraction as a result of temperature changes.

extrusive rocks a general term to encompass rocks of volcanic origin that are discharged onto the Earth's surface, e.g. LAVA flows.

eye the calm area at the centre of a hurricane.

eyot *see* AIT.

F

factors of production those things that are required to produce goods. They are usually classed as capital, labour and land. The capital includes buildings, plant and machinery (*fixed capital*) and components and raw materials (*circulating capital*).

fallow a term to describe agricultural ground that is left without a crop for a season to enable it to regain some of its fertility. Leaving a field to lie fallow is traditional practice in agriculture. It should not be confused with SET-ASIDE, which involves a payment from government in return for leaving a field uncultivated and ungrazed by cattle or sheep.

false colour a term used in REMOTE SENSING when using INFRARED PHOTOGRAPHY. Infrared radiation (heat) is invisible to the human eye but, by using special dyes on films, images of objects can be obtained from the infrared radiation that they emit.

false origin a point from which a grid is imposed on a map, chosen to prevent negative coordinates from appearing; i.e. the false origin will be to the south and west of the area to be mapped so that all EASTINGS and NORTHINGS are positive. In Britain, the false origin of the NATIONAL GRID is southwest of the Scilly Isles.

fan an area of detrital sediment occurring in a submarine environment at the base of cliffs or mountains. An *alluvial fan* is a sediment load deposited when a stream slows (and therefore loses its carrying power) caused by a decrease in gradient, e.g. on reaching a plain having flowed through a mountain range. A *cleavage fan* is where cleavage planes form fans in folded rocks.

fast ice a large area of sea ice joined to ice on the land. In summer it breaks up to form ICE FLOES and PACK ICE.

fathogram a chart of the sea floor obtained by taking echo soundings.

fathom a measurement of sea depth. 1 fathom = 6 feet (1.829 metres), 100 fathoms = 1 cable, and 1000 fathoms = 1 nautical mile.

fault *see* FOLDS AND FAULTS.

favela a South American term for a shanty town outside a city.

federalism a two-tier system of government designed to allow regions to have some control over their own affairs, e.g. education and planning, while the central government is responsible for defence, foreign policy and other matters affecting the whole country. The USA has a federal government.

fell an open mountainside of moors and rough grazing. 'Fell' occurs in many place names in the Lake District and the Pennines, probably evidence of Viking settlement.

fen a waterlogged area where the water is not acid, as in a PEAT BOG. A fen occurs in CHALK or CLAY areas. Patches of open water are fringed with alders, rushes and reeds.

fiard *see* FJARD.

field of view a REMOTE SENSING term for the angle through which an instrument can sense electromagnetic radiation.

filling the movement of air into a lower pressure area so that the DEPRESSION dies away.

fiord *see* FJORD.

firn a compacted snow that forms an intermediate between snow and glacial ice and has survived the melting of a summer.

firth a Scottish term for a narrow arm of the sea, most usually to describe an ESTUARY, e.g. the Firth of Clyde, but also to describe STRAITS, e.g. the Pentland Firth.

fissure a long, narrow crack in rocks.

fjard *or* **fiard** a sea inlet between low banks occurring on a drowned coastline. Fiards are found on the southern coast of Sweden. They are deeper than RIAS but do not have the deep, glaciated troughs of a FJORD.

fjord *or* **fiord** a long, deep, narrow inlet of the sea into a glaciated valley bounded by high, steep, rocky mountains. A distinctive feature of fjords is that they are shallower at the seaward end. This is thought to be the result of the fanning out of the GLACIER as it was released from the narrow valley onto the lowland.

flood plain the planar or near-planar surface at the bottom of a RIVER valley within which the river flows and which is covered when the river floods. It is formed by the progressive development of the river as it meanders laterally. It is made up of river-borne sediment (alluvium), which often shows a fining-upwards sequence from bedrock, through coarse gravels, then sand, followed by CLAYS and silts. This rhythmic sequence of sediments may be repeated.

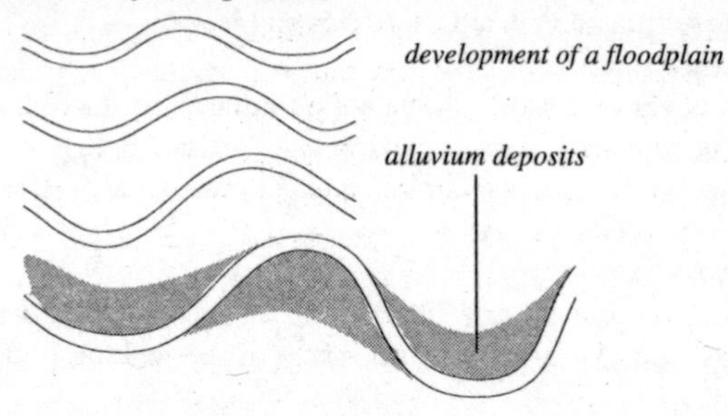

flood tide the incoming CURRENT of sea after low water. *See* EBB TIDE.

Florida current an important warm OCEAN CURRENT of the north Atlantic. It flows rapidly from the Caribbean through the Florida Straits to the Atlantic where it forms the southern part of the GULF STREAM.

fluvial pertaining to rivers.

focus the point of origin of an EARTHQUAKE.

foehn *or* **föhn** a general term for a warm, dry wind in the lee of a mountain range.

fog a suspension of water droplets (up to 20µm diameter) in the atmosphere when the air is near to saturation, causing visibility to fall below 1 kilometre (3280 feet). Fog formation is enhanced by the presence of smoke particles, which act as nuclei for the condensation of the droplets. The condensation may be caused by cooling of the ground or warm air moving over cold ground or water.

folds and faults geological features developed through tectonic activity. *Folds* are produced when rock layers undergo compression resulting in a buckling and folding of the rocks—a *ductile* or flowing deformation. There is an almost infinite variety in the shape, size and orientation of folds, and an earlier generation of folds may be refolded by subsequent periods of deformation. Their size may range from microscopic to folds occupying hillsides or even whole mountains.

The way folds are formed will depend on the amount of compression, the rock types involved and the thicknesses of the layers, because different rocks respond in different ways, but each fold possesses certain common features. The zone of greatest curvature is the *hinge*, and the *limbs* lie between hinges or on either side of the hinge. The *axial plane* is an imaginary feature bisecting the angle between limbs, and the *fold axis* is where the axial plane meets the hinge zone.

Faults are generally planar features, and are caused by *brittle* deformation. Rocks are moved, or *displaced*, across faults by as little as a few millimetres or as much as several kilometres (although possibly not all in the same event). There are several types of faults, depending on the movement across the fault and the orientation of the fault plane. The measurement of movement uses a horizontal and vertical component, the

two examples of vertical dip-slip fault movement (vertical section)

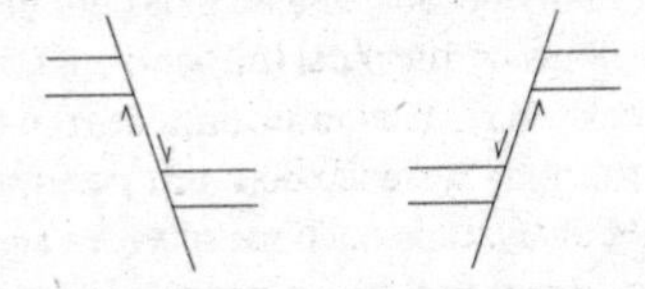

two examples of horizontal strike-slip fault movement (plan view)

heave and *throw* respectively. If the movement is up or down the fault plane, it is a *dip-slip fault*; if it is sideways, the term is *strike-slip*.

Many faults produce associated features such as zones of crushed rock, or striations (grooves) known as *slickensides*, on adjacent rock surfaces, where minerals such as quartz or calcite may grow.

Major faults today account for many surface features and are an important part of the PLATE TECTONIC structure and development of the Earth.

ford a point where a river or stream is shallow enough to be crossed without a bridge by people, animals or vehicles.

foredeep *see* DEEP.

fore-dune on a beach, the DUNE or line of dunes nearest the sea.

foreland a promontory of land jutting out from a coastline.

foreshock an EARTH TREMOR that occurs shortly before an EARTHQUAKE.

foreshore that part of a beach between the high and low water marks of SPRING TIDES.

form-line a line drawn on a map to give an impression of the terrain. There is not sufficient information to draw an accurate contour line, probably because the area has not been surveyed.

fossil fuels natural gas, PETROLEUM (oil) and coal, which are the major fuel sources today. They are formed from the bodies of aquatic organisms that were buried and compressed on the bottoms of seas and swamps millions of years ago. Over time, bacterial decay and pressure converted this organic matter into fuel.

Hard coal, which is estimated to contain over 80 per cent carbon, is the oldest variety and was laid down up to 250 million years ago. Another, younger variety (*bituminous coal*) is estimated to contain between 45 and 65 per cent carbon. The fuel values of coal are rated according to the energy liberated on combustion. Coal deposits occur in all the world's major continents, and some of the leading producer countries are the United States, China, Russia, Poland and the United Kingdom.

Natural gas consists of a mixture of hydrocarbons, including methane (85 per cent), ethane (about 10 per cent) and propane (about 3 per cent). However, other compounds and elements may also be present, such as carbon dioxide, hydrogen sulphide, nitrogen and oxygen. Very often, natural gas is found in association with petroleum deposits. Natural gas occurs on every continent, the major reserves being found in Russia, the United States, Algeria, Canada and in countries of the Middle East.

Petroleum is an oil consisting of a mixture of hydrocarbons and some other elements (e.g. sulphur and nitrogen). It is called crude oil before it is refined. This is done by a process called *fractional distillation*, which produces four major fractions:

(1) refinery gas, which is used both as a fuel and for making other chemicals;
(2) gasoline, which is used for motor fuels and for making chemicals;
(3) kerosine (paraffin oil), which is used for jet aircraft, for domestic heating and can be further refined to produce motor fuels;
(4) diesel oil (gas oil), which is used to fuel diesel engines.

The known residues of petroleum of commercial importance are found in Saudi Arabia, Russia, China, Kuwait, Iran, Iraq, Mexico, the United States and a few other countries.

Together, the fossil fuels account for nearly 90 per cent of the energy consumed in the United States. As coal supplies are present in abundance compared with natural gas or petroleum, much research has gone into developing commercial methods for the production of liquid and gaseous fuels from coal.

free market a market governed by supply and demand and with no government interference.

free port a port where no customs duties are payable, leading to lower insurance, administration and other costs.

free trade trade that takes place between countries without restrictions. It differs from a COMMON MARKET, in that member states each make their own policy on trade with non-members.

front the boundary between large masses of air with different properties. Fronts can be identified by the air masses separated at the front, their stage of development and the direction of their advance (*see* COLD FRONT and WARM FRONT).

frontier the limit of the settled area of a country. It may differ from the political boundary of the country.

frost frozen moisture. Air at less than freezing point, 0°C, forms air frost. GROUND FROST occurs when temperatures at ground level fall to freezing point. Frost is responsible for much EROSION, as moisture in cracks expands as it freezes, so widening the cracks or shattering the rocks.

fumarole a vent in a volcanically active area from which is emitted high-temperature gases (often H_2O, CO_2 and SO_2 in the main). It may indicate a late stage in volcanic activity, although it can precede eruptions.

gale a wind of force 8 on the BEAUFORT SCALE (more than 30 KNOTS).

gallery forest a strip of forest along RIVER banks in an otherwise treeless landscape.

garden city a planned settlement of low housing density with plentiful gardens and parks. Once the city reached a maximum size—about 30,000 inhabitants—no further growth would be allowed. Letchworth, followed by Welwyn Garden City, were Britain's first garden cities.

garden suburb suburbs with low housing density and plentiful open spaces.

GATT abbreviation of General Agreement on Tariffs and Trade, an agreement between free world countries to abolish trade barriers gradually.

GDP abbreviation of Gross Domestic Product, a measure of the total value of goods produced and services provided by a country in a year.

geest a heathland landscape on glacial sand and gravel, common in northern Germany.

gentrification the renewal of depressed inner city areas as poorer inhabitants move out and wealthier families buy the previously rented accommodation for redevelopment as housing. The area gradually becomes a fashionable place to live.

geo a narrow inlet in a sea cliff, worn by marine EROSION along a weak line in the rock.

geodesy essentially a combination of mathematics and surveying, involving the measurement of the shape of the Earth (or large areas of it).

Geographic Information System (GIS) a store of geographical data on a computer. It is a relatively recent development of cartography on computer and has grown rapidly with the advent of ever more powerful computers and sophisticated software. In such a system, every item of information has a geographic location and vast amounts of data can be retrieved, stored, analysed, displayed and used in a rapidly increasing number of professions and disciplines.

The information can be built up into a composite representation that includes the land and vegetation, land use patterns, rivers and streams, towns, cities and villages with all roads and modes of transportation, i.e. rail, bus routes, and so on. It is even possible to show all the public utilities, such as gas pipes, water mains, drainage, cables, etc. The uses of such an interactive database are vast, and many organizations use them. Demographic information can also be included to study development of towns, the need for facilities and other aspects of planning. It is even possible for the emergency services to use them to locate premises and utilities. The system will continue to develop, with greater capacity and flexibility and an increasing number of users.

geographics a computer printout perspective model of the terrain produced from grid-based map data.

geography the study of the Earth's surface, including all the land forms, their formation and associated processes, which comprise *physical geography*. Such aspects as CLIMATE, topography and oceanography are covered. *Human geography* deals with the social and political perspectives of the subject, including populations and their distribution. In addition, geography may cover the distribution and exploitation of natural resources, map-making and REMOTE SENSING.

geological timescale a division of time since the formation of the Earth (4600 million years ago) into units, during which rock sequences were deposited, deformed and eroded, and life of diverse types emerged, flourished and, often, ceased. The tables on pages 69 and 70 show the various subdivisions, ages, and appearance of flora and fauna. Many of the names owe their derivation to particular locations, rock sequences, and so on. The Roman name for Wales, *Cambria*, for example led to Cambrian, while the names of two Celtic tribes, the *Silures*

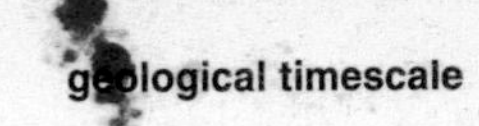

Major life forms over the geological record

Era	*Epoch*	*Life form*
Cenozoic	Recent	Modern Man
	Pleistocene	Stone Age Man
	Pliocene	many mammals, elephants
	Oligocene	pig and ape ancestors
	Eocene	
	Palaeocene	horse ancestors; appearance of cattle and elephants
Mesozoic	Cretaceous	end of the dinosaurs and the ammonites
	Jurassic	appearance of birds and mammals
	Triassic	dinosaurs appear; corals of modern type
Palaeozoic	Permian	amphibians and reptiles more common; conifer trees appear
	Carboniferous	coal forests; reptiles appear, winged insects
	Devonian	amphibians appear; fishes become more common; ammonites appear; early trees; spiders
	Silurian	first coral reefs; appearance of spore-bearing land plants
	Ordovician	trilobites and graptolites common; first fish-like vertebrates
	Cambrian	first period when fossils became common—trilobites, graptolites, molluscs, crinoids, radiolaria, etc.
	Precambrian	sponges, worms, algae and bacteria; all primitive forms

Oldest known signs of life are algae and bacteria about 3000 million years ago

Geological time scale

Eon		Era	Sub-era	Period	Epoch	*Millions of years since the start*
Phanerozoic		Cenozoic	Quaternary	Pleistogene	Holocene	0.01
					Pleistocene	2.00
			Tertiary	Neogene	Pliocene	5.10
					Miocene	24.60
				Palaeogene	Oligocene	38.00
					Eocene	55.00
					Palaeocene	65.00
		Mesozoic		Cretaceous		144.00
				Jurassic		213.00
				Triassic		248.00
		Palaeozoic	Upper Palaeozoic	Permian		286.00
				Carboniferous		360.00
				Devonian		408.00
			Lower Palaeozoic	Silurian		438.00
				Ordovician		505.00
				Cambrian		590.00
Proterozoic	PRECAMBRIAN					2500.00
Archaean						4000.00
Priscoan						4600.00

and *Ordovices*, provided the remaining Lo[illegible] names. Carboniferous is related to the prolifer[illegible] (i.e. carbon) and Cretaceous comes from *creta*, mea[illegible] 'chalk'. The Triassic is a threefold division in Germany, while the *Jura* mountains lent their name to the Jurassic. Cenozoic means 'recent life', Mesozoic, 'medieval life', Palaeozoic, 'ancient life', and Archaean, 'primeval'.

geology the scientific study of planet Earth. This includes geochemistry, petrology, mineralogy, geophysics, palaeontology, stratigraphy, physical and economic geology.

geomagnetism the study of the Earth's magnetic (geomagnetic) field, which has varied with time. At mid-oceanic ridges (*see* PLATE TECTONICS), where new crust is created, measurement of the geomagnetic field shows stripes, relating to reversals of the Earth's magnetic field, that are taken up in the newly formed rocks. This provides a tool in determining the age of much of the oceanic crust and is a vital piece of evidence in supporting the theory of plate tectonics.

geomagnetic fields of the mid-Atlantic ridge

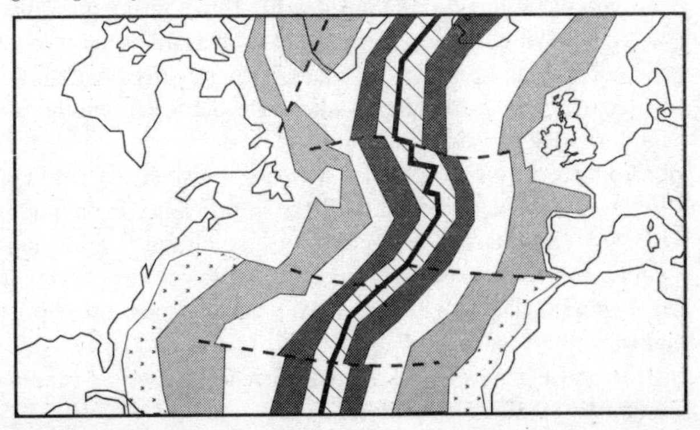

[illegible] a subject that grew out of GEOLOGY around the [illegible]e 19th century, and is the study of *land forms*, their [illegible]d change, i.e. the study of the Earth's surface. Land [illegible]s are made up of various ROCK types and formed from the s[illegible]face materials of the Earth by geomorphological processes that originate from tectonic movements and the CLIMATE. Land forms can be arranged into certain categories based on factors such as the underlying structural geology, the nature of the TOPOGRAPHY, and the terrain (soil, vegetation, etc), and the type of geomorphological processes dominant.

geostationary satellite a satellite that appears stationary to observers on the Earth because its velocity in orbit matches that of the Earth. They are used for communication, weather forecasting and REMOTE SENSING.

geyser a small fissure or opening in the Earth's surface, connected to a hot spring at depth, from which a column of boiling water and steam is ejected periodically. The mechanism of eruption is created by hot rocks heating water to boiling point at the base of the column before the top. Vapour bubbles rise through the column with considerable force, pushing out water and steam at the top. This reduces pressure at the base of the column, and boiling continues. Minerals dissolved in the water (calcium carbonate or silica) are deposited around the mouth of the geyser.

ghetto a part of a city occupied almost exclusively by a minority group whose way of life differs significantly from that of the host nation. The somewhat hostile attitude of the host populace tend to encourage the minority to continue to live in their ghetto. The original ghettoes were occupied by Jews in medieval Europe.

gill *or* **ghyll** a north of England term for a swiftly running mountain stream and its VALLEY.

GIS abbreviation of GEOGRAPHIC INFORMATION SYSTEM.

glacial action all processes related to the action of a glacier, including accumulation of crushed rock fragments and the physical actions, e.g. grinding, scouring and polishing, which are all due to the incorporation of rocks into the ice.

glacial deposits all deposits formed by some action of GLACIERS, e.g. boulder CLAY, sands and gravels occurring as outwash FANS, and deposits in the form of DRUMLINS and ESKERS.

glacial erosion the removal and wearing down of rock by glaciers and associated streams (of meltwater).

glacial lake a small LAKE dammed by ice between the edge of a GLACIER and the side of its VALLEY or MORAINE.

glacial outburst the sudden release of water from a glacial LAKE. If a large amount of water is released, the edge of the GLACIER may be lifted, so allowing the water to escape at great speed. Sometimes the lake water may escape through an opening in the glacier. Following the outburst, vast quantities of materials are carried down, causing great damage and completely changing the appearance of the VALLEY.

glaciated valley a U-SHAPED VALLEY with steep sides and a wide floor gouged out by GLACIAL EROSION.

glaciation a term meaning ice age, with all its effects, processes and products. The most recent is associated with the Pleistocene, but the rock record indicates older glaciations from the Precambrian and Permo-Carboniferous, and other periods in geological history.

glacier an enormous mass of ice, on land, that moves very slowly. About 10 per cent of the Earth's land is covered by glaciers, although during the last GLACIATION this was nearer 30 per cent. Glaciers that cover vast areas of land, e.g. Greenland, Antarctica, are called *ice sheets* (or *ice caps* if smaller), and these hide the underlying land features. The more typical glaciers are either those that flow in VALLEYS or those filling hollows in mountains. Glaciers can be classified further into *polar* (e.g. Greenland), *subpolar* (e.g. Spitzbergen) and *temperate* (e.g. the Alps) depending on the temperature of the ice.

Although glaciers move slowly, they act as powerful agents of EROSION on the underlying rocks. Large blocks may be dragged off the underlying rock, become embedded in the ice and then scratch and scour the surface as the glacier moves. This produces smaller particles of rock debris, and the blocks themselves may be broken. Debris is also gathered from valley sides and carried along. Ridges or piles of this rock debris are called MORAINE and, depending on position relative to the glacier, may mark present or former edges of the ice. In addition to the formation of MORAINE and associated characteristic formations, glaciers produce some typical large-scale features such as U-SHAPED VALLEYS and *truncated spurs*.

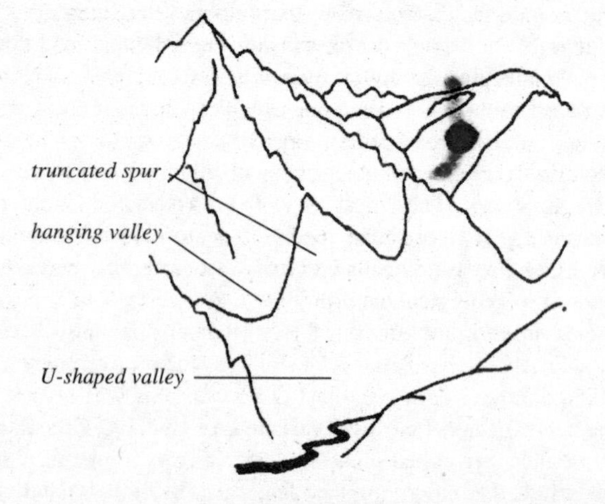

glen in Scotland, a steep-sided valley, narrower than a STRATH.

gley soils a mottled, blue-grey CLAY soil formed when the soil becomes waterlogged, thus excluding air. Iron in the soil is changed to its ferrous form, giving the characteristic blue-grey colour. It is organically rich.

global warming *see* GREENHOUSE EFFECT.

gloup a Scottish term for a BLOWHOLE.

GMT the abbreviation of Greenwich Mean Time, the time at longitude 0°, which passes through Greenwich in London. Britain uses GMT during the winter months and BST (British Summer Time) during the summer months. BST is one hour ahead of GMT. Worldwide, time is calculated from GMT.

GNP abbreviation of Gross National Product, which is calculated as the GDP of a nation plus income from investment abroad and minus income earned within the nation by foreigners.

Gondwanaland the massive hypothetical continent in the southern hemisphere that gave rise to parts of the present Africa, South America, India, Australia, New Zealand and Antarctica. Their connection at one time is postulated as a reason for the occurrence of widely separated but similar groups of plants and animals.

gorge *see* CANYON.

gradient a measurement of the steepness of a slope. It is the height ground rises or falls vertically in a given horizontal distance. It may be expressed in three ways: in the diagram, the

the gradient of AB is 1 in 4 or 25 per cent

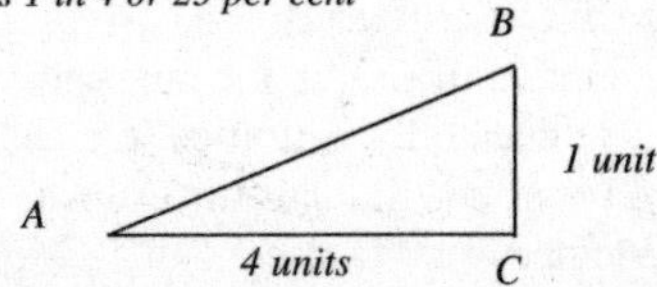

gradient of AC is (a) 1 in 4; (b) as a percentage $^{1}/_{4}$ x 100 = 25 per cent; (c) as an angle: < CAB = 14°. It is most often expressed as a percentage.

granite a coarse-grained and commonly occurring IGNEOUS ROCK containing quartz, alkali FELDSPAR and mica (usually biotite). Other minerals may also be present, including amphiboles and oxide minerals. Granites can be formed by several processes, including melting of old CONTINENTAL CRUST and fractional crystallization of basalt magma.

grasslands one of the four major types of vegetation, the others being FOREST, SAVANNA and DESERT. Grasslands are characterized by seasonal DROUGHT, limited PRECIPITATION and occasional fires, and these all, with grazing by animals, restrict the growth of trees and shrubs. Typical grasslands include the pampas of Argentina, the veldts of South Africa, the Steppes in Asia and the prairies of central North America.

great circle an imaginary circle on the Earth's surface the plane of which passes through the centre of the Earth. An arc of a great circle is the shortest distance betwen two places. This fact is important to airlines when planning routes, e.g. a flight from London to Los Angeles passes over Greenland.

greenhouse effect the phenomenon whereby the Earth's surface is warmed by solar radiation. Most of the solar radiation from the Sun is absorbed by the Earth's surface, which in turn re-emits it as infrared radiation. However, this radiation becomes trapped in the Earth's atmosphere by carbon dioxide (CO_2), water vapour and ozone, as well as by clouds, and is reradiated back to Earth, causing a rise in global temperature. The concentration of CO_2 in the atmosphere is rising steadily because of humankind's activities (e.g. deforestation and the burning of FOSSIL FUELS), and it is estimated that it will cause the global temperature to rise 1.5–4.5°C in the next 50 years.

Such a rise in temperature would be enough to melt a significant amount of polar and other ice, causing the sea level to rise by perhaps as much as a few metres. This could have disastrous consequences for coastal areas, in particular major port cities like New York.

Greenwich Mean Time *see* GMT.

grid plan an urban area where equidistant parallel streets are crossed at right angles by other equidistant parallel streets. The layout of New York is a good example.

grid reference a grid of horizontal and vertical lines on a map, numbered to show EASTINGS and NORTHINGS. Any point on the map can be pinpointed with reference to this grid.

Gross Domestic Product *see* GDP.

ground fog a FOG in low-lying areas and VALLEY floors. If the ground cools quickly at night, the lowest air layers are cooled. These cold layers collect in hollows and, if they are cooled to DEW POINT, a fog forms. Ground fog is usually only a metre or so in height with the atmosphere above being clear.

ground frost frost that forms when the minimum temperature at ground level is less than 0°C, although the air temperature above may be higher than this.

ground information a term used in REMOTE SENSING referring to information obtained from data on the physical state of the Earth. It includes maps, measurements of BIOMASS, temperature, soil moisture content, etc.

ground swell long, high waves whose tops are unbroken by the wind. Ground swell is almost always present in the open OCEAN.

groundwater water that is contained in the voids within rocks, i.e. in pores, cracks and other cavities and spaces. It often excludes *vadose* water, which occurs between the WATER TABLE and the surface. Most groundwater originates from the surface, percolating through the soil (*meteoric water*). Other sources are *juvenile* water, generated during and coming from deep magmatic processes, and *connate* water, which is water trapped in a sedimentary rock since its formation.

Groundwater is a necessary component of most weathering processes and of course its relationship to the geology, water table and surface may lead to the occurrence of AQUIFERS and artesian wells.

groyne a barrier built on a beach, running out towards the sea, to prevent the loss of beach material by LONGSHORE DRIFT.

guano accumulated bird droppings. It is rich in phosphates and nitrogenous material and is therefore an important source of fertilizer.

gulch a term used mainly in America for a deep GORGE.

gulf a huge inlet of the sea into a coastline. It is bigger than a BAY, e.g. the Gulf of Mexico.

Gulf Stream an important warm OCEAN CURRENT, originating in the Gulf of Mexico and flowing swiftly northwards, as the FLORIDA CURRENT, along the United States coast to about 40°N. It then flows northeastwards across the Atlantic Ocean. As it nears Britain and Norway, it becomes known as the *North Atlantic Drift* and is responsible for the temperate climate of northwest Europe and for keeping the west coast of Norway ice-free during winter.

gully a small, steep-sided RAVINE. It is formed by water cutting down through loose soil or rock in areas where vegetation has been lost. Gullies tend to be dry between periods of rainfall.

gust a sudden, short-lived increase of wind speed that may be associated with heavy rain as downdraughts sweep down from CUMULONIMBUS clouds. Gusts are more likely to occur in built-up areas than in open country where there is nothing to confine the wind. Gusts do more damage than steady winds.

gut a narrow channel where a RIVER joins an ESTUARY or the sea.

H

haar a cold FOG that occurs in spring and summer on the east coast of northern Britain. During a hot spell the ground heats up and the air immediately above is warmed and rises by CONVECTION. If that air moves out across the sea, the lower layers will be cooled by the sea and a fog may form. That fog, in turn, moves in across the land to replace the warm air that is rising from the land convectionally.

habitat the place where a plant or animal normally lives, specified by particular features, e.g. RIVERS, ponds, sea shore.

hachures short lines on a map that indicate the direction and steepness of a slope (although steepness is not quantified) but give no indication of altitude.

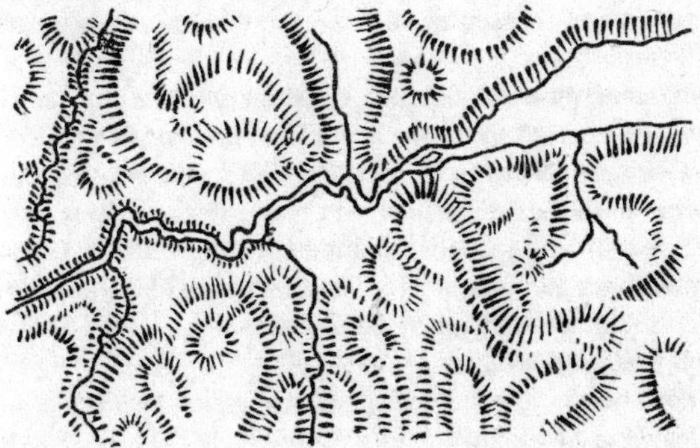

hamada a flat, bare, rocky DESERT plateau left when the wind has blown away all loose material. Although the remaining rock is polished by the wind, there is also some angular material caused by rocks splitting as they expand and contract in response to extremely hot day and cold night temperatures.

hanging valley a tributary valley situated above a major valley, possibly with a waterfall. It is formed by the greater EROSION of the main (trunk) valley by its GLACIER. The main glacier also cuts off the ends of the land between adjacent hanging valleys, creating truncated spurs (*see* illustration for glacier).

hard pan a hard, water-resistant layer in the soil formed after soluble minerals have been LEACHED out. The remaining insoluble material tends to get cemented together.

haze fine particles of salt spray, dust, smoke or other pollutant in the air that hinder visibility. Officially haze occurs when visibility is more than 1 kilometre but less than 2 kilometres.

headland (1) a high piece of land, usually with cliffs, jutting out into the sea. (2) the unploughed land at the edges of fields where ploughs turn.

head wall the steep back wall of a CIRQUE.

head waters the source of a RIVER at the head or top of its CATCHMENT area.

heat island the area of higher temperature over a city caused by the reflection of the Sun's heat from buildings and streets and by the heat generated by industry, traffic, etc. Heat islands develop during calm weather and may substantially reduce the number of AIR FROSTS or the number of snowfalls in a city.

hemisphere half of a sphere. The Earth is divided into the north and south hemispheres by the EQUATOR.

high pressure area *see* ANTICYCLONE.

hinterland the area behind a coastline. This also refers to the area served by a town or city.

hoar frost ice crystals formed on surfaces, e.g. vegetation, cooled by radiative heat loss. The ice comes from frozen DEW and the sublimation of water vapour to ice.

hog's back a steep, narrow ridge with symmetrical slopes. It forms where the DIP is almost VERTICAL.

holm a small islet, probably too small to be inhabited.

horn a pyramidal peak on a mountain left by a number of back to back CIRQUES that have cut back towards each other.

horse latitudes high pressure areas of calms and light winds lying 30°–35° north, and also south, of the EQUATOR.

hot deserts deserts found on the west coasts of land masses in tropical and subtropical latitudes. Average temperatures are greater than 25°C (77°F) and annual rainfall is less than 250 millimetres (10 inches). Vegetation is adapted to conserve water and may be spiny to discourage animals from eating it. Insects, reptiles, birds and small animals that inhabit hot deserts are all highly adapted to withstand drought and high temperatures.

hot spring a spring with a temperature of more than 37°C (98°F) that flows continuously out of the ground. It should not be confused with a GEYSER. Although it is more common in areas of volcanic activity, it can also occur in other areas. *See* MINERAL SPRINGS.

Humboldt Current *or* **Peru Current** a cold, nutrient-rich OCEAN CURRENT that flows northwards along the west coast of South America to the coast of Peru until it meets the EL NIÑO current. The Humboldt Current has a cooling effect on the coastal climate of Chile and Peru.

humidity the amount of water vapour in the Earth's atmosphere. The actual mass of water vapour per unit volume of air is known as *absolute humidity* and is usually given in kilograms per cubic metre (kgm^{-3}). However, it is useful to use *relative humidity*, which is the ratio, as a percentage, of the mass of water vapour per unit volume of air to the mass of water vapour per unit volume of saturated air at the same temperature.

hummock a low mound that may be of earth, rock or ice.

humus the material in soil that results from the decomposition of animal and vegetable matter. It provides a source of nutrients for plants.

hurricane (1) on the BEAUFORT SCALE, a wind with a mean speed of 64 KNOTS or more (74 mph). (2) an intense tropical cyclonic storm in which the winds circulate at high speeds. Hurricanes occur in the Pacific and North Atlantic. In the western Pacific, a hurricane is known as a *typhoon*.

husbandry the farming of animals.

hydration the addition of water to minerals. As a mineral takes up water, it swells, causing stresses in the rock. This is believed to be the main cause of crumbling in coarse-grained rock and is an important agent in both mechanical and chemical WEATHERING of rocks.

hydraulic action the force that the water in a RIVER or stream exerts on rocks in its path. A swiftly flowing CURRENT can remove loose material from the banks of a stream or river. In the case of a turbulently flowing current, the suction caused by EDDIES can lift material from the bed of the river.

hydroelectricity electricity that is produced when running water drives generators. The amount of electricity produced depends on both the mass of water and the height through which it falls. It is a popular method of producing electricity in mountainous areas, such as the Scottish Highlands or Norway.

hydrography the surveying and mapping of the sea bed, coastlines, OCEAN CURRENTS and TIDES. The resulting information is presented on CHARTS.

hydrology the study of water and its cycle, which covers bodies of water and how they change. All physical forms of water—rain, snow, surface water—are included, as are such aspects as distribution and use. The way in which water circulates between bodies of water such as seas, the atmosphere and the Earth forms the *hydrological cycle*.

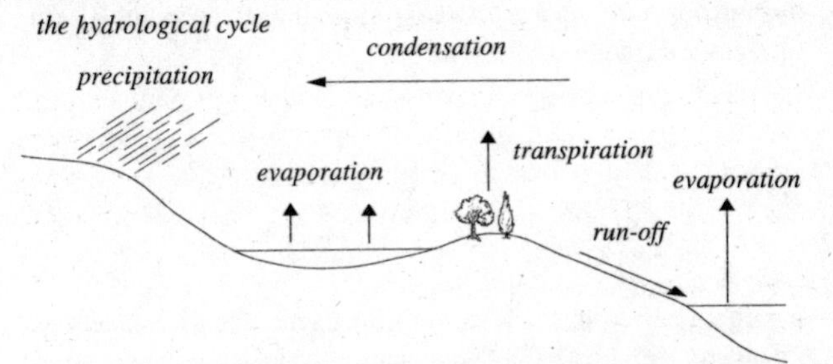

The cycle consists of various stages: water falls as rain or snow, of which some runs off into streams and then into lakes or rivers, while some percolates into the ground. Plants and trees take up water and lose it by transpiration to the atmosphere, while evaporation occurs from bodies of water. The water vapour in the air then condenses to cloud, which eventually repeats the cycle.

Water on the Earth

Of the 1.5 billion cubic kilometres of water
—the oceans hold 93.9 per cent
—groundwater 4.4 per cent
—polar ice 1.0 per cent
—lakes and rivers, the atmosphere and soil the remainder

hydrosphere the water that exists at or near the Earth's surface, whether as OCEANS, seas, LAKES, RIVERS or streams.

hygrometer an instrument for measuring the relative HUMIDITY of the air.

hypermarket an enormous retail development offering a wide range of goods. Hypermarkets are usually situated on the outskirts of towns or cities where there is ample parking space.

While this relieves traffic congestion in city centres, it can also lead to URBAN DECAY as smaller shops lose business.

hythergraph a graph of monthly rainfall against monthly temperature throughout the year.

I

ice age a period in the history of the Earth when ice sheets expanded over areas that were normally ice-free. The term is usually applied to the most recent episode in the Pleistocene (*see* GEOLOGICAL TIMESCALE), but the rock record indicates that there have been ice ages as far back as the Precambrian. Within ice ages there are fluctuations in temperature, producing interglacial stages when the temperatures increase.

ice barrier (1) a dam holding back the meltwaters of a glacier. (2) the edge of the ANTARCTIC ICE SHEET.

iceberg a large body of ice, derived from a GLACIER or ICE SHEET, floating in the sea. Icebergs can be carried long distances before finally melting.

ice cap ice covering an area of land, usually smaller in area than an ICE SHEET. However, it is used to refer to the Greenland ice cap, which is very large.

ice field in Canada, a term used unofficially to describe any large area of land ice. It also refers officially elsewhere to describe a wide area of ICE FLOES.

ice floe a large piece of flat, floating ice, smaller than an ICEBERG, formed when spring gales break up sea ice.

ice sheet a continuous sheet of thick ice, larger than an ICE CAP, that covers a land mass.

igneous rocks one of the three main rock types are the igneous rocks. They are formed by the intrusion of MAGMA at depth in various physical forms or extrusion at the surface as LAVA flows. Typical rocks are granite, basalt and dolerite.

Rocks erupted at the surface as lava are called *volcanic*, while *plutonic* rocks are those large bodies solidifying at some depth; *hypabyssal* rocks are smaller and form at shallow depths. When a body of magma has time to crystallize slowly, large mineral crystals can develop, while extrusion at the surface leads to a rapid cooling and formation of very small crystals (or none, if molten rock contacts water, as in a glass). A further division into *acid*, *intermediate*, *basic* and *ultrabasic* rocks is based on the silica (SiO_2) content, this being greatest in acid rocks.

Igneous intrusions may occur in several forms, either parallel to or cutting through the existing rocks, and the commonest are *sills* (concordant) and *dykes* (discordant). *Plugs* commonly represent the neck of a former volcano, while *batholiths* are massive, elongated bodies that may be hundreds of kilometres long.

examples of igneous intrusions

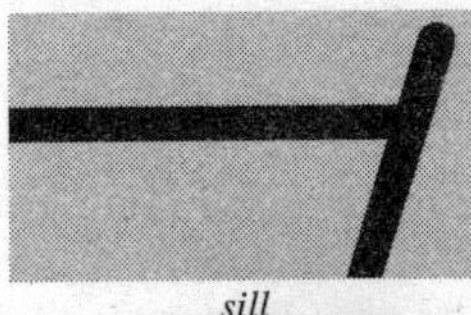

sill

dyke

batholith

plug (much smaller than batholith)

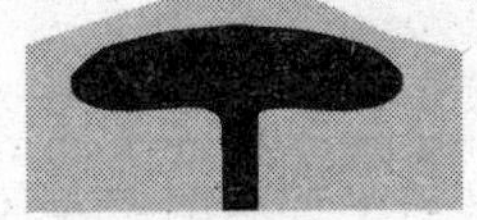

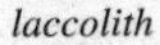

laccolith

Indian summer a period of warm, calm, dry, sunny weather sometimes occurring in late autumn in Britain.

indigenous a term used to describe a plant or animal native to a region.

industrial crop a non-food crop grown as a raw material for industry, e.g. conifers are grown for wood pulp, and linseed oil and linen are made from flax.

industrial inertia the survival of an industry in a location after the reasons for establishing it there no longer apply. For example, it is usually cheaper to update a factory or other enterprise than to try to move a skilled workforce to a new location.

infrared photography a REMOTE SENSING technique used, for example, in surveying vegetation, estimating BIOMASS or in night photography. It makes use of the fact that all living things emit heat, or infrared radiation. The infrared radiation is captured on special film, and its intensity can be used to provide an image of the subject being photographed. *See* FALSE COLOUR.

infrared radiation electromagnetic radiation with wavelengths between those of visible light and microwaves, i.e. from about 0.75 μm to 1 millimetre. Such radiation can penetrate fog and, by using special photographic plates, details invisible to the naked eye may be rendered visible (*see also* ELECTROMAGNETIC WAVES).

infrastructure the basic system of roads, railways, power supplies, communications, etc, needed for economic development. Efficient economies require highly developed infrastructures.

inselberg a steep hill, usually of granite or some other resistant rock, rising abruptly from a plain. It is typical in tropical landscapes. It comes from German, meaning 'island mountain'.

insolation the Sun's radiation reaching the Earth that depends, *inter alia*, on the orbital position of the Earth and the translucency of the atmosphere.

insular a term pertaining to islands.

interfluve an area of ground separating two RIVERS that both run into the same DRAINAGE system.

International Date Line an imaginary but internationally agreed line drawn along the 180° MERIDIAN, to the east of which the calendar date is one day earlier than to the west. It deviates slightly to east and west to avoid crossing land. When moving eastwards across the line, a traveller repeats a day and, similarly, loses a day when moving westwards across it. This arrangement allows for the accumulated time change of 1 hour for every 15° of longitude east or west of the Greenwich meridian. (*See* LATITUDE AND LONGITUDE).

intertropical convergence zone a narrow zone at low latitude where AIR MASSES from north and south of the EQUATOR converge, often resulting in DEPRESSIONS, which may lead in the ocean areas to HURRICANES when the zone is displaced from the Equator.

inversion of temperature a situation where the usual temperature gradient in the atmosphere is reversed and temperature increases vertically. This occurs often in anticyclonic conditions, producing stable air near the ground on clear nights.

invisible exports services rather than goods, such as insurance, advertising or banking, that are sold abroad and therefore earn foreign currency.

ionosphere the layer of the atmosphere beyond 80 kilometres (50 miles) above the surface of the Earth. It can reflect radiomagnetic waves back to Earth and is therefore important for radio communications. Transmission may be affected by magnetic storms or sunspot activity, both of which disturb the ionosphere.

irrigation the supply of water to dry land to encourage or facilitate plant growth. The water may be applied by means of canals, ditches, sprinklers or the flooding of the whole area. Flood irrigation is not always a good idea because the water evaporates from the flooded field, leaving behind any salts that were dissolved in the water. If this is done repeatedly, the build-up of salts may harm the soil and make it infertile. Water may be conserved and used more effectively by means of a pipe with small holes being laid around a plant, the water dripping from the holes onto the soil. Irrigation can transform ARID regions, but RIVERS from which water is diverted will inevitably be much reduced.

island arc a line of VOLCANOES on the continental side of a deep oceanic trench that marks the subduction of oceanic crust. Almost three-quarters of past or present volcanoes are in a belt around the Pacific, especially along the volcanic island arcs. EARTHQUAKES are associated with these areas, because of the downwards movement of the subducted slab, and the gradual meeting of the plate at depth releases magma and fluids that rise to generate the volcanism of the island arcs.

isobar a line used to join points of equal atmospheric pressure on a weather map at a given time. If there is a great change in pressure over a small area then the change in weather is more apparent, and this is shown on a weather map by closely drawn isobars.

isohel a line joining locations that have the same amount of sunshine.

isohyet a line joining locations that have the same rainfall.

isotherm a line drawn on a map joining places of equal temperature.

isthmus a narrow neck of land connecting two larger pieces of land, e.g. the Isthmus of Panama.

J

jet stream westerly winds at high altitudes (above 12 kilometres/7½ miles), found mainly between the poles and the TROPICS, that form narrow jet-like streams. The air streams move north and south of their general trend in surges, which are probably the cause of DEPRESSIONS and ANTICYCLONES. There are a number of separate jet streams, but the most constant is that of the subtropics. Jet-stream speed and location is of importance to high-flying aircraft.

joint a fracture in rock that may occur as a parallel set or, more commonly, in an irregular and less systematic manner. Where a set of joints can be identified, it can usually be related to tectonic stresses and the geometry of the rock body. There are several types of joint: *unloading joints*, which are caused by release of stress on rocks at depth as overlying rocks are removed by EROSION; *cooling joints*, which occur in igneous bodies; and joints related to regional deformation.

jungle a general term for more or less impenetrable tropical forest.

juvenile water water that originates from a MAGMA and has never been in the atmosphere. Surprisingly, water in great quantities can originate in this way.

K

kame a structure produced by glacial deposition (*see* GLACIAL DEPOSITS). It occurs as a hummock of sands and gravels, with bedding, and often exhibiting slumping at the sides. It was formed by the melting of stagnant ice, which caused the load to be dropped.

karst a karst landscape is one created in a limestone area and created by the limestone itself. The distinctive land forms are a result of dissolution by solutions that move through joints and fissures and dissolve calcareous material, enlarging cracks and joints and creating underground waterways and CAVES. Features typical of a karst include networks of furrows and sharp crests, funnel-shaped hollows, and ultimately conical hills and steep-sided DEPRESSIONS of impressive scale.

katabatic wind the sinking and downward movement of cold, dense air beneath warmer, lighter air. The air is cooled by radiation, usually at night. It occurs over ice-covered surfaces and in the FJORDS of Norway, and in many cases can be gale force.

kettle hole a hole or DEPRESSION formed in glacial drift as a result of outwash material from a GLACIER covering isolated masses of ice. When the covered ice eventually melts, the sediments slump down into the space, creating a surface depression.

key *see* CAY.

kibbutz a type of Israeli commune where all agricultural land and resources are jointly owned and all farming plans are made jointly.

knoll a low, rounded hill.

knot a unit of nautical speed equal to one nautical mile (1.15 statute miles or 1.85 kilometres) per hour. The term 'knot' originates from the period when sailors calculated their speed by using a rope with equally spaced tied knots attached to a heavy log trailing behind the ship. The regular space between knots was measured at 47 feet, 3 inches, which is 14.4 metres.

kopje the Afrikaans term for a small, isolated rocky hill.

Köppen classification a system of climatic classification that was developed between 1910 and 1936, and is based on annual and monthly measurements of temperature and PRECIPITATION, and the major types of vegetation. The system comprises three orders or levels, beginning with the *overall climate*, e.g. warm, temperate and rainy (class C), which can then be categorized further, e.g. 'winter, dry' produces class Cw. The third level qualifies temperature, e.g. a hot summer would produce a classification Cwa.

kyle the Gaelic term for the straits between two islands or an island and the mainland, e.g. the Kyles of Bute.

L

Labrador Current an important cold OCEAN CURRENT that flows southwards from the ARCTIC along the west coast of Greenland until it meets the warm NORTH ATLANTIC DRIFT on the Grand Banks, off Newfoundland. The mingling of cold and warm water helps to make the Grand Banks a rich fishing ground but is also responsible for the FOGS that are a feature of the area. The Labrador Current carries ICEBERGS and ICE FLOES south into the North Atlantic shipping routes.

lacustrine a term pertaining to LAKES.

lagoon (1) shallow salt water almost cut off from the sea by a beach or REEF. (2) the sheltered deep water within an ATOLL.

lahar a mudflow developed on the flank of a volcano under the combined effects of eruption and torrential rainstorms (or melting of ice or snow). The majority of volcanic fatalities are caused by lahars, which can travel many kilometres from the source. If an eruption interacts with a crater lake producing a hot mudflow, then the result can be even more catastrophic.

lake a large volume of standing inland water. Some lakes, e.g. the Great Salt Lake, are salty, but most are freshwater.

land and sea breezes air circulation along coasts during summer, developed when the overall pressure gradient is minimal. During the day the Sun warms the land more than the adjacent sea, and so the air above the land becomes warmer and thus rises. This produces a convective motion, with the cooler air from the sea moving onto the land. At night the situation is reversed because the sea is warmer than the land.

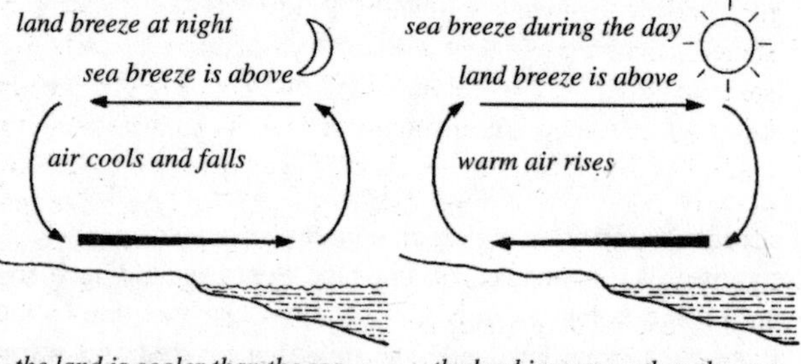

land bridge a land connection between continents along which human beings and animals could migrate. It is thought that at one time a land bridge existed between the northern part of Asia and Alaska.

land classification the classification of land according to its potential agricultural output. As well as soil quality, other factors such as elevation, rainfall, drainage, susceptibility to EROSION, etc, are taken into consideration.

LANDSAT one of the series of unmanned NASA satellites that orbits the Earth specifically to acquire data about the Earth. It orbits the Earth 14 times per day at an altitude of 900 kilometres (560 miles) and passes over each point of the Earth's surface once every 18 days. It records information using REMOTE SENSING techniques and transmits it to receiving stations on Earth.

landscape architecture the planning and management of land to fulfil functions in an aesthetically pleasing way. It may be applied to very diverse cases, such as motorways and parks.

latent heat the measurement of heat energy involved when a substance changes state. While the change of state is occurring, the gas, liquid or solid will remain at a constant temperature, independent of the quantity of heat applied to the substance (an increase in heat will just speed up the process). The specific latent heat of fusion is the heat needed to change one kilogram of a solid into its liquid state at the melting point for that solid. For example, the specific latent heat of fusion for pure, frozen water (ice) at 273K (0°C) is 334 $kJkg^{-1}$. The specific latent heat of vaporization is the heat needed to change one kilogram of the pure liquid to vapour at its boiling point. In the case of pure water again, at its boiling point of 373K (100°C), 2260$kJkg^{-1}$ is the specific latent heat of vaporization needed to change water into steam.

lateral erosion this occurs when a RIVER erodes its banks. It is most marked in a MEANDER, where the faster flow of water undercuts the bank on the outside of the meander. In time, as the process is repeated, the channel moves downstream.

lateral moraine rock debris created by a GLACIER, which accumulates at the margin of a valley glacier and is caused by transport and reworking of rocks from the valley sides, which eventually accumulate as MORAINE.

latitude and longitude latitude is the angular distance, north or south from the EQUATOR, of a point on the Earth's surface. The

Equator is 0°, and points can therefore be measured in degrees south or north of this line. The imaginary lines drawn on a map or globe are the *lines of latitude*.

Longitude is a similar concept. It is the angular distance of a point measured on the Earth's surface to the east or west of a 'central' reference point. The reference point in this case is the plane created by a MERIDIAN (an imaginary circle that cuts the poles and goes over the Earth's surface and the point in question) going through Greenwich in England. A point may be 0° longitude if it sits on this line or a number of degrees east or west. The Greenwich Meridian, based on the Greenwich Observatory, was established by an international agreement in 1884. There is a time difference of one hour when travelling 15° of longitude at the Equator (*see also* TROPICS).

lava molten rock erupted by a volcano, whether on the ground (*subaerial*) or on the sea floor (*submarine*). Both acidic and basic forms are found and show a number of textures, but all are characterized by some glass and/or fine-grained minerals caused by rapid cooling. The way it is erupted and moves, and therefore its subsequent morphology, depends greatly on the viscosity; generally, a less viscous lava will flow faster. Two types of basaltic lava forms are seen: *aa*, which is jagged, and *pahoehoe*, which exhibits a smoother, ropy appearance.

law of the sea *see* SEA, LAW OF THE.

leaching the percolation of water down through the soil, dissolving out humus and soluble salts and depositing them in the underlying layers of soil. The upper layers become acidic, minerally deficient and less fertile.

leat an artificial channel to carry water to a reservoir or for some industrial purpose, e.g. turning a mill wheel or washing ore.

lee *or* **leeward** the side of a hill, building or other object that is protected from the wind.

lee shore the shore towards which the wind is blowing.

legend the key to the symbols on a map.

levanter a strong easterly wind of the western Mediterranean that blows occasionally between July and October and usually brings rain and storms.

leveche a hot, dry, dust-laden wind blowing from the Sahara into Spain. It blows ahead of the DEPRESSIONS that move from west to east through the Mediterranean during the early summer months.

levee a river bank that has been built up above the level of the surrounding FLOOD PLAIN. When a RIVER overflows its banks and floods surrounding fields, the water on the fields flows much more slowly than the main stream. Any material in this flood water will be dumped, so raising the river banks. Naturally formed levees tend to raise the level of the river bed above the level of the surrounding countryside. If the levee is breached, the resulting flooding is widespread and serious since water pours onto the surrounding area instead of flowing along the course of the river. A graphic example of such flooding occurred in the spring of 1996 when the Mississippi breached its levees. Levees can also be manmade.

lightning the discharge of high voltage electricity between a CLOUD and its base and between the base of the cloud and the Earth. (It has been shown that in a cumulonimbus cloud positive charge collects at the top and negative at the base). Lightning occurs when the increasing charge (of electricity) in the cloud overcomes the resistance of the air, leading to a *discharge*, seen as a flash. The discharge to ground is actually followed by a return discharge up to the cloud, and this is the visible sign of lightning. There are various forms of lightning, including *sheet*, *fork* and *ball*, and it may carry a charge of around 10,000 amps.

The *thunder* is the rumbling noise that accompanies lightning and it is caused by the sudden heating and expansion of the air by the discharge, causing sound waves. The sound often continues for some time because sound is generated at various points along the discharge—the latter can be several kilometres long. The thunder comes after the lightning because sound travels more slowly than light, and this allows an approximate measure of distance from the flash to be made. For every five seconds between the flash and the thunder, the lightning will be roughly one mile away.

limestone a sedimentary ROCK that is composed mainly of calcite with dolomite. Limestone may be organic, chemical or detrital in origin. There is a tremendous variety in the make-up of limestones, which may comprise remains of marine organisms (corals, shells, etc); minute organic remains (as in CHALK); or grains formed as layered pellets in shallow marine water and also calcareous muds. Recent deposits of calcium carbonate are found in shallow tropical seas. Modification after deposition is usually extensive, and both the composition and structure may change because of compaction, recrystallization and replacement.

linear development *see* RIBBON DEVELOPMENT.

lithosphere that layer of the EARTH, above the ASTHENOSPHERE, that includes the crust and the top part of the mantle down to 80–120 kilometres (50–75 miles) in the OCEANS and around 150 kilometres (93 miles) in the continents. The base is gradational and varies in position depending on the tectonic and volcanic activity of the region. The lithosphere comprises the blocks that constitute PLATE TECTONICS.

Little Ice Age a period of climatic cooling between about 1550 and 1850. GLACIERS in the northern HEMISPHERE advanced several kilometres down their VALLEYS and temperatures were much colder than at present. The NORTH ATLANTIC DRIFT was pushed farther south, crossing to Spain and North Africa rather than to Britain and Norway. The reasons for the Little Ice Age are not understood, but it is thought that it may have been something to do with volcanic eruptions that occurred about that time. Vast clouds of dust and gas blotted out much of the Sun's radiation. It is also possible that CLIMATE may be influenced by the long-term cycles of sunspot activity. Relatively few sunspots were seen in the latter part of the 17th century, while much more sunspot activity was seen during the warm period of 1000 years ago.

littoral a term pertaining to a shoreline, whether sea shore or the shore of a lake.

littoral zone the area between high and low water marks during ordinary spring TIDES. The definition is sometimes extended to include water to a depth of 200 metres. In this case, water between 60 and 200 metres (197 and 656 feet) deep is called the *sublittoral zone*.

loam a good, fertile, easily worked soil made up of about 40 per cent SAND, 40 per cent SILT and 20 per cent CLAY.

loch the Scottish term for a LAKE or for a long narrow sea inlet.

lochan a small, freshwater LOCH, similar to a CIRQUE lake.

lode a vein or fissure in a rock containing mineral deposits.

loess a sediment formed from the aeolian transportation and deposition of mainly silt-sized particles of QUARTZ. It is well sorted, unstratified and highly porous, and although it can maintain steep to vertical slopes, is readily reworked. Loess is widespread geographically, and although thicknesses of Chinese deposits exceed 300 metres (984 feet), it is normally a few metres. Its origin has been hotly debated, but it is now accepted that the loess particles are produced by glacial grinding, frost cracking, hydration in desert regions and aeolian impact of sand grains. The wind is the primary and essential agent in the process.

longitude *see* LATITUDE AND LONGITUDE.

longshore drift the movement of material (sand and shingle) along the shore by a current parallel to the shoreline (i.e. a *longshore current*). Longshore drift occurs in two zones: *beach drift*, at the upper limit of wave activity (i.e. where a wave breaks on the beach); and the *breaker zone* (where waves collapse in shallow water) where material in suspension is carried by currents.

lough the Irish term for a LAKE or narrow sea inlet.

lunar pertaining to the Moon. A *lunar day*, 24 hours 50 minutes, is the time taken by the Moon to make two successive crossings of the same MERIDIAN on Earth. This is why high TIDE is 50 minutes later on each successive day.

lunar month the time between new moons, approximately 29 days.

M

machair the white shell sand found in western Scotland, especially in the Western Isles. It forms a light, easily worked soil. Where it is not cultivated, it produces rich grassland.

mackerel sky a cloud pattern comprising wavy CIRRO- or ALTOCUMULUS with holes, suggesting the markings of a mackerel.

macroclimate the general CLIMATE of a large area, which may be a country or may be as large as a continent.

macro-economics the study of the large-scale economic issues, such as fiscal policy, that affect the economy of a country as a whole.

magma the fluid rock beneath the Earth's surface that solidifies to form IGNEOUS ROCKS. During volcanic eruptions, the LAVA extruded at the Earth's surface is not necessarily the same in composition as the magma that arises to form lava, since the magma may have lost some of its gaseous elements and some solids of the magma may have crystallized.

magnetic declination *or* **magnetic variation** the acute angle between magnetic north and geographic (true) north in degrees east or west of true north. It varies in different parts of the world according to the relative positions of the true and magnetic north poles to that of the observer.

magnetic deviation the angle through which a COMPASS has to be corrected to compensate for the effects of MAGNETIC FIELDS in its immediate surroundings, e.g. those fields that may be set up by the ironwork in a ship.

magnetic dip *or* **magnetic inclination** the angle at which lines of magnetic force caused by the Earth's magnetic field appear to enter the Earth's surface. The angle of magnetic dip is 0° at the EQUATOR and 90° at the MAGNETIC POLES.

magnetic field the region of space in which a magnetic body exerts its force. Magnetic fields are produced by moving charged particles, and they represent a force with a definite direction. There is a magnetic field covering all of the Earth's surface, which is believed to be a result of the iron-nickel core.

magnetic meridian an imaginary line joining the positions of the MAGNETIC POLES. A freely swinging COMPASS needle will align itself along a magnetic meridian as long as there is no interference from other MAGNETIC FIELDS in the area.

magnetic poles the points where the lines of magnetic force of the Earth's MAGNETIC FIELD are vertical. The magnetic poles do not coincide with the geographic north and south poles and their positions slowly change. During the 1980s the magnetic north pole was at 70°N, 100°W and the magnetic south pole was at 68°S, 143°E. A line drawn through the Earth to join them misses the centre of the Earth by approximately 1200 kilometres.

M and G are the positions of the magnetic and geographic north poles respectively. O_1 *and* O_2 *are the positions of observers.*
M, G and O_1 *are in a straight line. Therefore the magnetic declination at* O_1 *is zero. At any other position, e.g.* O_2, *magnetic declination is between 0° and 90° and is given by the angle* MO_2G

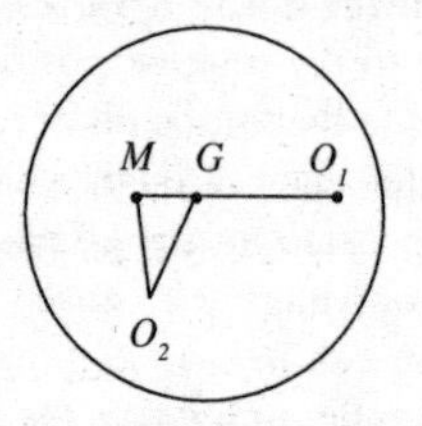

magnetic variation *see* MAGNETIC DECLINATION.

map projection a system for depicting the features of the Earth, the land masses and OCEANS, onto a flat sheet of paper. Because the Earth is spherical, all projections introduce some distortion. *Mercator's projection*, which is frequently used in atlases, distorts and enlarges land areas towards the poles; *azimuthal projections* show true direction; *orthographic projections* attempt to give the effect of a globe; *gnomic projections* show the shortest distance, in a straight line, between two places.

map reference *see* GRID REFERENCE.

maquis the French word that has been generally adopted to describe the low, evergreen, shrub vegetation that is characteristic of the northern coast of the Mediterranean Sea. It includes many aromatic herbs and other sun-loving plants that are happy to grow on rather poor soils.

marginal land poor quality land that is difficult to cultivate and unlikely to be profitable. It is usually brought into cultivation only in times of hardship. At other times it provides rough grazing.

maritime a term pertaining to the sea.

maritime climate the climate of areas in mid-latitudes that is moderated by the influence of the sea. Winters tend to be cool and wet, summers warm and moist. Maritime climates are found predominantly on the west coasts of continents as well as on islands.

marram grass one of the commonest forms of vegetation found on sand DUNES. Its roots spread over a wide area, both horizontally and vertically, which makes it an ideal agent for stabilizing a dune.

marsh an area of low-lying, poorly drained ground, often containing small stretches of open water. In temperate latitudes, reeds, rushes and other rough grasses grow in the marsh with, possibly, some alder and willow trees. Trees are more common in marshes in TROPICAL areas. When marshes are drained, they usually reveal a very fertile area of ground, because of their high organic content, e.g. the Fens in England.

massif a clearly defined mountainous area that has a uniform appearance and geology and is distinct from the surrounding area.

meander the side-to-side wandering of a stream or RIVER channel, best developed in river deposits on the FLOOD PLAIN. The origin of meanders is not established absolutely but involves the original stream course, the natural hydrodynamic properties of water flowing over sediment or rock, and the fact that the river takes a course that requires the least energy to follow. Once established, meanders create EROSION and deposition on the outer and inner banks respectively, and the curved river form can become more exaggerated and 'migrate' downstream. A variety of features may be developed, including OXBOW LAKES.

mean sea level (MSL) the average level of the surface of the sea, calculated from measurements taken over a long period of time. British mean sea level is based on data from Newlyn in Cornwall and is used as the ordnance datum in ORDNANCE SURVEY maps.

Mediterranean climate hot, dry summers and warm, wet winters. This type of CLIMATE is experienced on the western edges of continents between latitudes 30° and 40° north and south of the EQUATOR. LAND AND SEA BREEZES influence daily weather, while local winds have a big seasonal effect; e.g. the MISTRAL, LEVANT and SIROCCO. Subtropical ANTICYCLONES ensure fine weather during the summers, while DEPRESSIONS, bringing rain, are common during the winters.

megalopolis any CONURBATION of more than 10 million inhabitants.

Mercator projection *see* MAP PROJECTION.

mere a small, shallow LAKE. The word is common in Cheshire and Shropshire.

meridian an imaginary circle on the Earth's surface connecting the geographic poles. The PRIME MERIDIAN has a value of 0° and passes through Greenwich. It is also known as the Greenwich Meridian. Other meridians are known by the angle they make, east or west, with the prime meridian.

mesa a steep-sided tableland, larger than a BUTTE. It is found where layers of sedimentary STRATA have a hard, resistant top layer, which may be a harder sedimentary rock, or an igneous sill, or LAVA flow. It is found in arid or semi-arid areas and is thought to be a stage in the denudation of a PLATEAU.

metamorphic rocks rocks formed by the alteration or

recrystallization of existing rocks by the application of heat, pressure, change in volatiles (gases and liquids), or a combination of these factors. There are several categories of *metamorphism* based on the conditions of origin: *regional*—high pressure and temperature as found in *orogenic* (mountain-building) areas; *contact*—where the rocks are adjacent to an igneous body and have been altered by the heat (with little or no pressure); *dynamic*—very high, confined pressure with some heat, as generated in an area of faulting or thrusting, i.e. where rock masses slide against each other; *burial*—which involves high pressure and low temperature, as found, for example, at great depth in sequences of sedimentary rocks.

The key feature of all metamorphic rocks is that the existing *assemblage* (group) of minerals is changed by the pressure and/or heat and the presence of fluids or other volatiles. New minerals grow that are characteristic of the new conditions. Some typical metamorphic rocks are schist, slate, gneiss, marble, quartzite and hornfels. Depending on the type of metamorphism, there are systems of classification into *zones* or *grades* where specific minerals appear in response to increasing pressure and/or temperature.

meteorology the scientific study of the conditions and processes within the Earth's atmosphere. This includes the pressure, temperature, wind speed, cloud formations, etc, that, over a period of time, enable meteorologists to predict likely future WEATHER patterns. Information is generated by weather stations and also by satellites in orbit around the Earth.

metropolis a large city together with its suburbs. There are no size limits to either area or population. *See also* MEGALOPOLIS; CONURBATION.

microclimate the climate peculiar to a small area of, perhaps, only a few square centimetres.

mid-oceanic ridge throughout the world's oceans are long, linear volcanic ridges, in effect submarine mountain chains, generally positioned centrally. The ridges are sites where new oceanic crust is created through the spreading of the plates (a spreading axis) and outpouring of basalt. The mid-oceanic ridges are sites of shallow EARTHQUAKES, and they are often cut by transform faults that create an offset appearance to the ridge. Those faults linking the ridge are also the focus of earthquakes.

mile (1) the statute mile measures 1.609 kilometres (1760 yards). (2) the geographical mile, which is the distance of 1 MINUTE along the EQUATOR (1.852 kilometres). (3) the British nautical mile, which measures 1 minute of arc at latitude 48°. Other nations define the nautical mile as 1 minute of arc at 45°. One British nautical mile equals 1.00064 international nautical miles.

millibar a unit of measurement of ATMOSPHERIC PRESSURE equal to one thousandth of a BAR. One millibar is a pressure of 1000 dynes per cm^2 or 100 Newtons per m^2.

mineral springs springs that produce water that contains a high concentration of dissolved mineral salts and therefore thought to be of therapeutic value. Frequently, the water is warm or hot (*see* HOT SPRINGS). The 1980s and 1990s saw a big increase in the number of firms bottling mineral spring water for sale. SPA TOWNS often develop in an area with mineral springs.

minute (1) in time, one minute is one sixtieth of an hour. (2) in distance one minute is one sixtieth of a degree of LATITUDE or longitude.

mirage a visual phenomenon caused by the reflection and refraction of light. A mirage is seen wherever there is calm air with varying temperatures near the Earth's surface. A common mirage in the desert is caused by the refraction of a downward light ray from the sky, so that it seems to come from the sand, and, to any onlooker, it would appear that the sky is reflected in a pool of water. As the inverted image of a distant tree is also usually formed, the overall effect resembles a tree being reflected in the surrounding water.

misfit stream *see* UNDERFIT STREAM.

mist water droplets in suspension, which reduce visibility to not less than 1 kilometre (1093 yards). *See also* FOG.

mistral the cold, dry winter wind that blows strongly down the Rhône VALLEY from the Alps to the Mediterranean Sea. It has a marked cooling effect on the CLIMATE of the area.

mixed cultivation the growing of two crops in a field, either so that one can shade the other or in an attempt to reduce soil EROSION. The system is more likely to be practised in TROPICAL rather than temperate areas.

mixed economy in contrast to a FREE MARKET economy, one in which there is considerable government intervention. This may be during production or distribution of goods and services. For example, EC governments intervene in food production by providing subsidies or imposing QUOTAS.

mixed farming the type of agriculture where stock rearing and crop production take place on the same farm. Much of the crop is used as animal fodder, and the animal manure is spread on the fields, so cutting down the need for artificial fertilizers.

monoculture the growing of a single crop by a farmer. The advantages are increased efficiency and quality control. The disadvantages are that the farmer cannot spread his or her risks with regard to CLIMATE and market conditions, and the crop becomes more susceptible to disease as those organisms that attack it build up in the soil year after year.

monopoly the domination of the market for goods or services by a single supplier. In practice, monopolies never occur.

monsoon in general terms, winds that blow in opposite directions during different seasons of the year, with features that are associated with widespread temperature changes over land and water in the subtropics. Monsoon winds are essentially similar in origin to LAND AND SEA BREEZES but occur on a very much larger scale geographically and temporally. The Indian subcontinent is subjected to a rainy season in the southwesterly monsoon. Other areas affected by monsoons are Asia (east and southeast), parts of the West African coast and northern Australia.

Moon the EARTH's one satellite, at an average distance from Earth of 384,000 kilometres (238,600 miles). It has no ATMOSPHERE, water or magnetic field, and surface temperatures reach extremes of 127°C (261°F) and –173°C (–279°F). The surface is heavily cratered, probably because of meteorite impact, and a distinctive feature of the Moon is its maria, which are huge dry plains on its surface that were once thought to be seas. Rock collected on Apollo space missions has indicated an age of about 4000 million years. A feldspathic crust up to 120 kilometres (75 miles) in thickness overlies a mantle of silicates, and basaltic (*see* BASALT) LAVAS cover almost one fifth of the surface. There is probably a small core of iron, with a radius of approximately 300 kilometres (186 miles).

moor open upland of rough grass, heather and bracken, often on peaty soil. It can provide rough grazing but has no other agricultural value.

moraine a general term for ridges of rock debris that were deposited by GLACIERS and mark present or former ice margins, although it originally referred to ridges of debris around alpine glaciers. There are numerous forms of moraine, from LATERAL MORAINE to *terminal moraine* (deposited at the leading edge of an active glacier). A *medial moraine* is a merging of lateral moraines as a result of the convergence of two glaciers. There are essentially two types of moraine when considering formation. *Dump moraine* is where material is literally dumped (during glacier retreat) at the margin or end of a glacier when the ice is stationary. *Push moraines* denote the edges of ice advance when the ice moves over sediment and pushes up ridges. Large moraines of this type can be seen, often formed annually, and stratified sediments may be faulted and folded because of the force of the glacier.

mountain a hill more than 2000 feet (600 metres) in height. The formation of mountain chains clearly involves phenomenal movements of the Earth's crust and unimaginable forces, even although the process takes place over many millions of years. The process of mountain building (called OROGENY or *orogenesis*) involves the accumulation of enormous thicknesses of sediments, which are subsequently folded, faulted and thrusted, with igneous intrusions at depth, producing rock complexes involving sedimentary, igneous and metamorphic ROCKS.

A massive linear area that has been compressed in this way is called an *orogenic belt*, and the formation of such belts is interpreted by means of PLATE TECTONICS. Different mechanisms are postulated for the formation of mountain chains, e.g the Andes by subduction of oceanic lithosphere; the collision of continents for the Himalayas, and the addition of vast basins of sedimentary rocks and ISLAND ARCS onto an existing plate in the case of the North American Cordillera (*see* PLATE TECTONICS for definition of terms).

The result of these global crustal movements is the mountain ranges as we see them today, where the higher peaks belong to the younger ranges. The highest points on the seven continents are as follows:

Peak	Height	Continent	Country
Everest	8848m	Asia	Nepal/China
Acongagura	6960m	South America	Argentina
McKinley	6194m	N. & Central America	Alaska
Kilimanjaro	5895m	Africa	Tanzania
Elbrus	5642m	Europe	CIS
Vinson Massif	5140m	Antarctica	Antarctica
Mauna Kea	4205m	Oceania	Hawaii

Mountains have a considerable effect on local weather conditions, and south-facing slopes in the northern hemisphere are warmer and drier than north-facing slopes because they receive more sun. When warm, moist air reaches a mountain, it cools as it rises, releasing moisture, so on the leeward side the air descends and absorbs moisture. In many instances, deserts occur on the leeward side of a mountain range, e.g. the Gobi Desert (Asia) and the Mojave Desert of western North America. It is well known that the temperature falls on ascending a mountain (about 6°C per 100 metres). Mountains therefore have different plants depending on the altitude.

MSL abbreviation of MEAN SEA LEVEL.

mull (1) a Scottish term for a headland. (2) the HUMUS produced by decomposition of grass or forest litter.

multiband camera a REMOTE SENSING camera used in aerial photography and specially adapted to expose different areas of film so that photographs of the same area are produced in different wavebands. Objects with different reflective abilities can then be distinguished from each other.

multispectral scanner in REMOTE SENSING, a SENSOR in a SATELLITE or aircraft that simultaneously records images of the Earth's surface in various wavebands.

multiplier a term used in economics to describe the increased effect of an action on an area's economy. If cash is injected into an area by, for example, building a factory and providing new jobs, the economy grows by more than the value of the original cash injection because of the growth in demand for services, e.g. housing or leisure pursuits. Similarly, the closure of a business leads to a higher loss to the economy than the value of the wages and salaries paid by that business, as the demand for services also drops, causing further redundancies in other businesses.

municipal a term pertaining to local government of a town or city.

N

nacreous clouds a cloud formation at great height before sunrise or after sunset when its colouring is similar to mother of pearl.

nadir (1) in REMOTE SENSING, the point on the ground vertically below the centre of the SENSOR. (2) the lowest point reached by an object in orbit.

national grid (1) the system of cables for supplying electricity to all parts of the country. (2) the reference system of coordinates on ORDNANCE SURVEY maps, illustrated overleaf.

National Park a large area of scenic land that is protected by a government body for the use and enjoyment of the people. It may contain SITES OF SCIENTIFIC INTEREST. Although development and industry are not wholly forbidden in a National Park, there are strict controls imposed. National Parks in Britain are found in the Lake District, the Peak District, Snowdonia and Exmoor, among other areas. In total, some 20 per cent of the land area of Britain is officially protected as National Parks.

The national grid as used on Ordnance Survey maps, showing the false origin

	0–100	100–200	200–300	300–400	400–500	500–600	600–700
900–	NA	NB	NC	ND	NE	OA	OB
800–900	NF	NG	NH	NJ	NK	OF	OG
700–800	NL	NM	NN	NO	NP	OL	OM
600–700	NQ	NR	NS	NT	NU	OQ	OR
500–600	NV	NW	NX	NY	NZ	OV	OW
400–500	SA	SB	SC	SD	SE	TA	TB
300–400	SF	SG	SH	SJ	SK	TF	TG
200–300	SL	SM	SN	SO	SP	TL	TM
100–200	SQ	SR	SS	ST	SU	TQ	TR
0–100	SV	SW	SX	SY	SZ	TV	TW

nautical mile the standard international unit of distance used in navigation. One nautical mile is defined as 1852 metres (6076 feet).

neap tide a tide of a range up to 30 per cent less than the mean tidal range, which occurs near the Moon's first and third quarters (every 14 days) when the Moon, Earth and Sun are at right angles and the Sun's tidal influence works against the Moon's.

needle a pointed spire of rock, detached from a MOUNTAIN rockface or sea cliff, e.g. the Needles off the Isle of Wight.

neritic zone the shallow water marine zone near the shore that extends from low tide to a depth of approximately 200 metres (656 feet). Most sea-floor organisms live in this zone because sunlight can penetrate to such depths. Sediments deposited here comprise sands and clays.

ness a word meaning HEADLAND. It occurs in many place names, e.g. Dungeness, Holderness, etc.

névé an alternative term for FIRN.

nimbus *or* **nimbostratus** a grey layer of cloud that obscures the Sun and produces continuously falling rain (or snow).

noctilucent clouds thin, very high clouds of ice or dust, which are sometimes brilliantly coloured as a result of their reflecting light from the Sun when it is below the horizon.

nomads people with no fixed abode who move from place to place with their animals seeking pasture. They may or may not follow the same routes each year. Many desert tribes and the Lapps of northern Scandinavia are nomads.

North Atlantic Drift *see* GULF STREAM.

Northern Lights *see* AURORA.

northing the second part of a GRID REFERENCE, always coming after the EASTING, when map coordinates are quoted. It represents the distance travelled northwards from the reference line of the grid.

North Pole the northern end of the Earth's AXIS. The geographic and MAGNETIC POLES rarely coincide.

nullah a watercourse in India that is filled after monsoons but dry at other times.

nunatak a rock MOUNTAIN peak that protrudes above the ice in an area undergoing GLACIATION.

nutation a small shift in the Earth's axis, occurring about every 19 years, resulting from the gravitational pull of the Sun and Moon on the Earth's equatorial bulge.

O

oasis a place in a DESERT that has sufficient water to support vegetation. Deep wells may have to be dug to reach the water, or it may rise naturally at an ARTESIAN WELL, or the WATER TABLE may be exposed in a DEFLATION HOLLOW. Oases vary in size and can cover a reasonably large area.

oblate a term meaning slightly flattened at the poles, as is the Earth.

occlusion the meeting of warm and cold air in a DEPRESSION, where the warm air is lifted above the colder air.

ocean and ocean currents technically, those bodies of water that occupy the ocean basins, the latter beginning at the edge of the continental shelf. Marginal seas such as the Mediterranean, Caribbean and Baltic are not classed as oceans. A more general definition is all the water on the Earth's surface, excluding lakes and inland seas. The oceans are the North and South Atlantic, North and South Pacific, Indian and Arctic. Together with all the seas, the salt water covers almost 71 per cent of the Earth's surface.

From the shore the land dips away gently in most cases—the continental shelf—after which the gradient increases on the continental slope leading to the deep sea platform (at about 4 kilometres depth). There are many areas of shallow seas on the continental shelf (*epicontinental seas*), e.g. North Sea, Baltic Sea and Hudson Bay. In the ice age, much of the shelf would have been land and, conversely, should much ice melt, the continents would be submerged further. The floors of the oceans display both mountains, in the form of the MID-OCEANIC

RIDGES, and deep trenches. The ridges rise 2–3 kilometres (1.25–1.8 miles) from the floor and extend for thousands of kilometres while the trenches reach over 11 kilometres (6.8 miles) below sea level at their deepest (the Mariana Trench, southeast of Japan).

Ocean zones

littoral zone		between low and high water spring tides
pelagic zone	0–180 m	floating plankton and swimming nekton
neritic zone	low tide–180 m	benthic organisms
bathyal zone	180–1800 m	beyond light penetration, but much benthic life (crawling, burrowing or fixed plants and animals)
abyssal zone	> 1800 m	
abyssal plain	~ 4000 m	ooze of calcareous and siliceous skeletal remains; red clay only below 5000 m

The oceans contain *currents*, i.e. faster moving large-scale

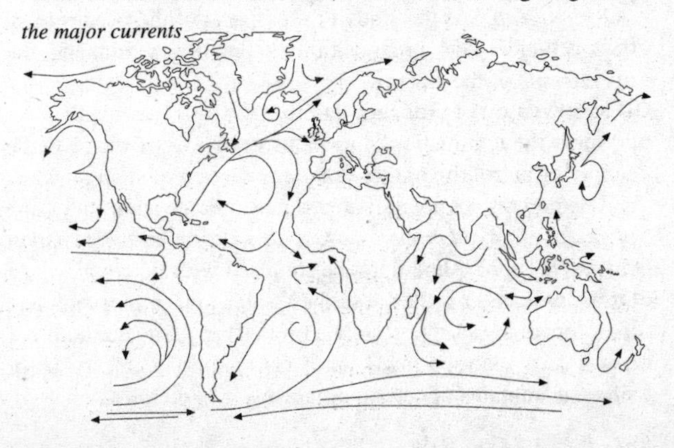

flows (the slower movements are called *drifts*). Several factors contribute to the formation of currents, namely the rotation of the Earth, prevailing winds, differences in temperature and sea water densities. Major currents move clockwise in the northern hemisphere and anticlockwise in the southern hemisphere. Well-known currents include the GULF STREAM and the HUMBOLDT CURRENT.

The composition of seawater (specified as ions present) is the result of, for all but 0.1 per cent, eleven components:

	ion	*parts per thousand*
Cl^-	chloride	19.0
Na^+	sodium	10.6
SO_4^{2-}	sulphate	2.6
Mg^{2+}	magnesium	1.3
Ca^{2+}	calcium	0.4
K^+	potassium	0.4
HCO_3^-	bicarbonate	0.1
Br^-	bromide	less than 0.1
F^-	fluoride and others	

Oceanography is the study of all aspects of the oceans from their structure and composition to the life within and the movements of the water.

OD abbreviation of ordnance datum. *See* MEAN SEA LEVEL.

oligopoly the domination of an industry by a few firms, e.g. the supply of petrol to garages or the production of soap powders and detergents. In practice, firms must charge about the same price for their products in order to hold on to their market share. Oligopoly should not be confused with MONOPOLY.

oligotrophic a term that describes water low in nutrients and therefore able to support only a limited range of aquatic life.

ooze a deep sea mud made up of CLAYS and the calcareous or siliceous remains of certain organisms, e.g. diatoms.

OPEC abbreviation of Organization of Petroleum Exporting Countries, the member countries of which are Algeria, Iran, Iraq, Kuwait, Libya, Qatar, Saudi Arabia, the United Arab Emirates, Nigeria, Gabon, Ecuador, Venezuela and Indonesia. They meet at regular intervals to fix the price of oil on world markets.

opencast mining a system of mining in which the top layers of ROCK are removed, the mineral ores removed by earth-moving machinery and the surface layers of rock and soil replaced. It is cheaper than deep mining using tunnels and mines.

opportunity cost a term used in economics to describe cost in terms of the sacrificed alternative. For example, the opportunity cost of conserving a scenic area may be the jobs that failed to materialize because a mining operation was refused planning permission. Similarly, the opportunity cost of a new motorway may be the loss of a scenic area.

orbit the path of a heavenly body (and satellite or spacecraft) moving around another because of gravitational attraction.

ordnance datum *see* MEAN SEA LEVEL.

Ordnance Survey (OS) in Britain, the government body responsible for carrying out surveys so that it can produce and publish maps.

ore any naturally occurring substance that contains commercially useful metals or other compounds. The extraction of the desired metal will only proceed if the process is both economically and chemically feasible. Some relatively unreactive metals, such as copper and gold, exist as native ores with no need of extraction, but most metals are obtained by extracting them from their oxygen-containing (oxide) ores.

orogeny a period of MOUNTAIN building, many of which have occurred in the past, each lasting for millions of years.

orographic rainfall *see* RELIEF RAINFALL.

OS the abbreviation of ORDNANCE SURVEY.

outback the remoter, rural parts of Australia.

outcrop exposure at the Earth's surface of part of a ROCK formation.

oxbow lake the development of river MEANDERS into large loops. Eventually the 'neck' between a looped meander is cut and the river straightens its course, leaving a cut-off loop or oxbow lake.

development of a river meander leading to the formation of an oxbow lake

oxygen a colourless and odourless gas that is essential for the respiration of most life forms. It is the most abundant of all the elements, forming 20 per cent by volume of the ATMOSPHERE; about 90 per cent by weight of water; and 50 per cent by weight of ROCKS in the crust.

ozone a denser form of oxygen that exists as three atoms per molecule (O_3). Ozone is a more reactive gas than the more common diatomic molecule (O_2), and can react with some hydrocarbons in the presence of sunlight to produce toxic substances that are irritants to the eyes, skin and lungs. Minute quantities of O_3 are found in sea water. It forms the Earth's ozone layer, 15 to 30 kilometres (9 to 18 miles) above the Earth's surface.

ozone layer a part of the Earth's ATMOSPHERE, at approximately

15–30 kilometres (9–19 miles) height, that contains ozone. Ozone is present in very small amounts (one to ten parts per million), but it fulfils a very important role by absorbing much of the Sun's ultraviolet radiation, which has harmful effects in excess, causing skin cancer and cataracts and unpredictable consequences to crops, and plankton.

Recent scientific studies have shown a thinning of the ozone layer over the last 20 years, with the appearance of a hole over Antarctica in 1985. This depletion has been caused mainly by the build-up of CFCs (chlorofluorocarbons) from aerosol can propellants, refrigerants and chemicals used in some manufacturing processes. The chlorine in CFCs reacts with ozone to form ordinary oxygen, lessening the effectiveness of the layer. CFCs are now being phased out, but the effects of their past use will affect the ozone layer for some time to come.

P

Pacific-type coast *see* CONCORDANT COAST.

pack ice any floating ice that is not attached to the land. It can include both ICEBERGS and ICE FLOES.

pampas the temperate grasslands of southern America. Much of the pampas is now farmland or used for cattle ranching. *See* PRAIRIE; STEPPES; VELD.

pantograph an instrument used to enlarge or reduce the scale of maps or drawings.

parent material the substrate from which and on which a soil is formed. The soil may be found on a rock surface that is weathered, or it may result from the WEATHERING of superficial deposits (glacial and river deposits).

passive glacier a glacier the rate of flow of which is extremely slow and the bulk of which is more or less constant. Although a small amount of ice is lost by summer melting, this is replaced by a small accumulation of SNOW during the winter. Passive glaciers occur in the interior of continents or ICE SHEETS where precipitation is low.

passive remote-sensing system a SENSOR that does not transmit electromagnetic radiation but detects radiation from the Earth's surface. *See* ACTIVE REMOTE-SENSING SYSTEM.

paternoster lakes a series of elongated lakes in a glacial valley, connected by a RIVER. The end result resembles the beads on a rosary, hence the name.

peat an organic deposit formed from compacted dead and possibly altered vegetation. It is formed from vegetation in swampy hollows and occurs when decomposition is slow because of the lack of oxygen in the waterlogged hollow. Sphagnum moss is among the principal source plants for peat. As peat builds up each year, water is squeezed out of the lower layers, causing the peat to shrink and consolidate. Even so, cut peat has a high moisture content and is dried in air before burning. In Ireland and Sweden, peat is used in power stations.

pediment a gentle slope at the foot of a MOUNTAIN that extends at a low gradient towards a RIVER. There is a marked change of angle between the steep slope above and the gradient of the pediment, which is always less than 10° and usually less than 6°. Pediments may be formed as slopes retreat or as running water deposits surface soil or debris.

pelagic a term used to describe organisms living in the sea between the surface and middle depths. Pelagic sediments (e.g. ooze) are deep-water deposits comprising minute organisms and small quantities of fine-grained debris.

peneplain the final stage in an erosional cycle when a region is worn down to an undulating plain characterized by low relief, with small hills and wide, shallow RIVER valleys. Residual hills may remain, and these are called *monadnocks*, after Mount Monadnock in New Hampshire, USA.

peninsula a piece of land that is almost an island. It is connected to the rest of the land mass by a narrow neck of land, e.g. the Gower Peninsula in Wales.

perched water table an AQUIFER that lies on top of impermeable ROCK, which is, in turn, above the level of the normal water table of the area.

percolation the process by which water moves downwards through soil and through the joints, cracks and pores in rocks. The quantity of water that moves in this way can be measured using a *percolation gauge*. It may be important to measure the rate of percolation to estimate how quickly substances may LEACH out of a soil.

perihelion the point in the orbit of a heavenly body when it is nearest the sun it circles. The Earth is at its nearest point to the Sun on 3 January, when the distance between them is 92 million miles or 147.3 million kilometres. Perihelion can apply to planets, comets, spacecraft, etc.

permafrost ground that is permanently frozen save for surface melting in the summer. About one quarter of the Earth's land surface is affected, and although it may be very thick (several hundred metres), the larger depths are probably relics from the last ICE AGE. It is technically defined as being when the temperature is below 0°C for two consecutive years, and it can extend to depths of several hundred metres. The top layer that thaws in the summer is called the *active layer*, and there may be unfrozen ground between this and the permafrost, a zone that is called *talik*. Depths of 1500 metres (4900 feet) in Siberia and 650 metres (2100 feet) in North America have been recorded, and today permafrost underlies 20–25 per cent of the Earth's land area—a figure that was much greater during parts of the Pleistocene.

The ground in areas of permafrost shows distinctive features, including patterns of circles, polygonal cracks, mounds and PINGOS. *Polygonal cracks* are the result of contraction caused by cooling in winter, and in Spitzbergen the polygons may reach 200 metres across. *Mounds* are caused simply by the increase in volume that accompanies freezing of water, which pushes up surface layers of soil. Large mounds (40–50 metres high) are called *pingos*.

Peru Current *see* HUMBOLDT CURRENT.

petroleum *or* **crude oil** a mixture of naturally occurring hydrocarbons formed by the decay of organic matter, which, under pressure and increased temperatures, forms oil. The often mentioned '*reservoir*' is the rock in which oil (and gas) is found, and common types of reservoir rock are sandstone, LIMESTONE or dolomite. The oil migrates, after formation, from the source rocks to the reservoir (because such vast quantities could not have been formed in place) where it must be contained by a *trap*. A trap is a particular geological configuration where the oil is confined by impermeable rocks.

typical geological traps

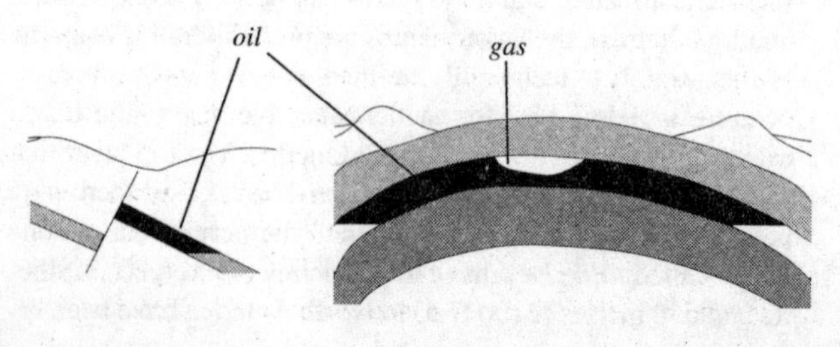

Most of the world's petroleum reserves are in the Middle East although the CIS and USA currently produce a significant proportion of the world's oil.

Major oil-producing countries

	% world share
CIS	18
USA	13
Saudi Arabia	10.5
Iran	5
Mexico	4.5
China	4.5
(UK	3.0)

Oil reserves

	% world share
Saudi Arabia	25
Iraq	10
Kuwait	9.5
Iran	9
Abu Dhabi	9
Venezuela	6
CIS	5.5
(UK	0.5)

The modern oil industry began over a century ago when a well was bored for water in Pennsylvania, and oil appeared. Petroleum also occurs in the form of ASPHALT or *bitumen*, syrupy liquids or near solid in form, and there are significant deposits today. The Pitch Lake of Trinidad, over 500 metres across and about 40 metres deep, is fed from beneath as the asphalt is removed. There are similar occurrences in Venezuela and California, and in Alberta, Canada, are the famous Athabasca Oil Sands, where the sandstone is full of tar, an oil of asphalt.

Petroleum consists of many hydrocarbons of differing composition, with small amounts of sulphur, oxygen and nitrogen. The components are separated and treated chemically to provide the basic building blocks and products for the vast petrochemicals industry.

pH index a scale for expressing how acidic or alkaline a solution is. It is widely used to describe the acidity or alkalinity of soils. Neutral soils have a pH value of 7; very acid, peaty soils have a pH value of between 3 and 4 and very alkaline soils have a pH value of between 10 and 11. The pH scale runs from 0 to 14 and is the measure of concentration of hydrogen ions (H^+) in an aqueous solution. The pH is the negative logarithm (base 10) of H^+ ion concentration, calculated using the following formula:

$$pH = \log_{10} (1/[H^+])$$

The scale of pH ranges from 1.0 (highly acidic), with decreasing acidity until pH 7.0 (neutral) and then increasing alkalinity to 14 (highly alkaline). As the pH measurement is logarithmic, one unit of pH change is equivalent to a tenfold change in the concentration of H^+ ions.

photic zone the uppermost layer of a LAKE or sea where there is adequate light to allow photosynthesis to proceed. The limit will vary, depending on the quality of the water and the material held in suspension, but can be as much as 200 metres (656 feet).

photochemical fog a haze, often incorrectly called SMOG, produced when sunlight reacts with vehicle emissions in hot, dry, calm conditions. As there is no water vapour in the atmosphere, it is not a true FOG. However, visibility is poor. It has become a feature of some cities during calm, sunny weather and is a hazard to health, especially to people with respiratory problems. Athens and Los Angeles are two particularly badly affected cities, although most cities suffer to a greater or lesser extent.

photocontour map a map that shows the surface features of an area, obtained from information acquired by aerial photography.

photomap a quickly made map obtained by adding place names, boundary lines, etc, to a mosaic of aerial photographs. It does not normally show CONTOUR lines.

piedmont glacier an extension of ice from a valley GLACIER, which projects beyond its valley walls onto the adjacent flat plain at the foot of the MOUNTAINS (i.e. the piedmont). Because the ice is now at a lower altitude, it may show more rapid diminution as a result of melting.

pillow lava subaqueous, basaltic LAVA flows are characterized by pillow structure, each rarely more than one metre in diameter but often forming a sequence hundreds of metres thick. Pillow lavas form from long flow tubes, and as the sea water causes the rapid cooling of each lobe, so further branching and budding occurs to continue the outpouring.

pingo the Eskimo term for a dome-shaped hill with a core of ice. The hill may rise to 60 metres or more in height. There are two theories as to how they are formed. Firstly, in a PERMAFROST region, as all the water in a lake freezes, it can only expand upwards. As the ice rises, it carries the lake sediments and vegetation with it to form a dome shape. This is known as a *closed-system pingo*. Secondly, the *open-system pingo* occurs where ground water flows under ARTESIAN pressure under thin permafrost. As the water forces itself upwards, it freezes and pushes upwards as a core of ice. Any surface vegetation covers this cone in a dome shape.

pisciculture a scientific term for fish farming.

plateau a raised, extensive, relatively flat area of land. There is usually a steep slope on at least one side of the plateau, falling sharply to lower land. For example, the high VELD of South Africa forms a plateau separated from the low veld by the ESCARPMENT.

plate tectonics a concept that brings together the variety of features and processes of the Earth's crust and accounts for continental drift, sea-floor spreading, volcanic and earthquake activity, and crustal structure.

It has long been noticed how coastlines on opposite sides of oceans, e.g. the Atlantic, seemed to fit together. Other geological features led to the theory that continents were joined together millions of years ago. This theory was supported by a reconstruction of fossil magnetic poles and in 1962 by the idea of sea-floor spreading where ocean ridges were the site of new crust formation, with slabs of crust moving away from these central sites. All this was brought together with the idea that the lithosphere (the crust and uppermost part of the mantle) is made up of seven large and twelve smaller plates composed of oceanic or continental crust. The plates move relative to each other with linear regions of creation and destruction of the lithosphere.

There are three types of plate boundary:

ocean ridges	where plates are moving apart (constructive);
ocean trenches	where plates are moving together (also for young mountain ranges) (destructive);
transform faults	where plates move sideways past each other (conservative).

At destructive plate boundaries, one tectonic plate dips beneath the other at an oceanic trench in a process called *subduction*, and in so doing old lithosphere is returned to the mantle.

The effects of this are belts of earthquake and volcanic activity linked to the melting of the old plate at depth. ISLAND ARCS are an example of volcanic activity associated with subduction at an ocean trench, where there are very often also earthquakes. Where two continental plates converge, the continents collide to produce mountains as seen in the Alps and Himalayas today. The transform faults of conservative plate boundaries are generated by the relative motion of two plates alongside each other, and the best-known example of this is the San

Andreas fault in California, a region that suffers earthquakes along this major fracture.

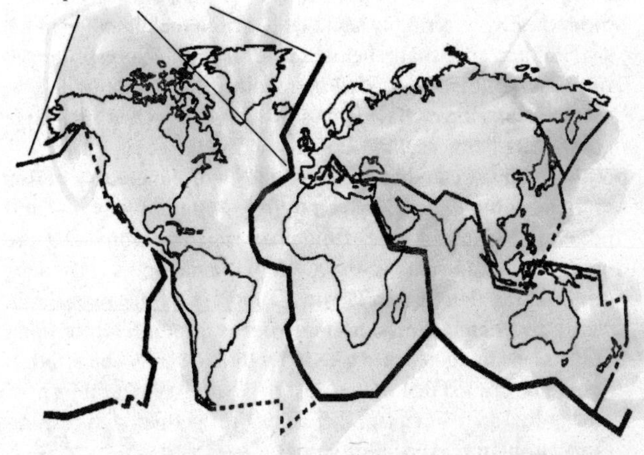

podsol *or* **podzol** a soil with minerals LEACHED from its surface layers into lower layers. Podsolization is an advanced stage of leaching, which involves the removal of iron and aluminium compounds, humus and clay minerals from the topmost horizons and their redeposition lower down.

polar air mass a mass of air that forms in latitudes 40°–60° north or south of the Equator. Where it forms over the ocean it is called a *maritime polar air mass* (symbol mP, *see* AIR MASS). If it forms over a continent it is a *continental polar air mass* (symbol cP). It is separated from TROPICAL AIR MASSES by polar FRONTS where DEPRESSIONS generated as cold and warm air meet.

polar high an area of semi-permanent high pressure (ANTICYCLONE) that forms over the NORTH and SOUTH POLES. The anticyclone over the ARCTIC is rather less permanent than that over the ANTARCTIC. As polar highs advance into mid latitudes, they bring colder weather with them.

polder the Dutch term for land that has been reclaimed from the sea. Polders are usually below sea level and must be protected from the sea by DYKES. They are drained by an elaborate system of canals through which water is pumped out to sea. Huge polders are being reclaimed from the Zuider Zee.

pollution an undesirable change in the environment that makes it an unhealthy place in which to live. Pollution may be natural, e.g. the ash and gases from a volcanic eruption, but more often pollution occurs as the result of humankind's activities.

Air pollution may be caused by burning FOSSIL FUELS (coal, oil, gas), by vehicle emissions (*see* PHOTOCHEMICAL FOG) or other emissions by industry, e.g. sulphur dioxide (*see* ACID RAIN). It could be argued that pollen dust in the air is a natural form of air pollution as it causes difficulties for people with respiratory problems, especially asthma.

Freshwater pollution is caused mainly by industry or farming. If excess fertilizer is applied, the nitrogen may run off into streams, rives and lakes. Silage effluent is another source of pollution from farming. When fresh water becomes heavily polluted, all plant and animal life dies and a serious hazard is posed to the quality of drinking water.

Sea-water pollution is caused mainly by oil spillages and sewage outfalls.

Ground pollution is caused by the often illegal disposal of chemicals by industry and may render sites unsuitable for development.

There are other forms of pollution, such as *noise pollution* and *light pollution*, the excessive use of bright lighting that blocks out the night sky.

pollution dome the mass of warm, polluted air that accumulates over a city in calm weather when there is a layer of colder air preventing it from rising (*see* HEAT ISLAND). If a wind then blows it away, it is known as a *pollution plume*.

prairie a vast, more or less flat, natural grassland found in mid latitudes. There are few trees. Rainfall is low, summer temperatures high, and winters are cold. It is excellent agricultural ground. The word was first applied to the prairies of Canada and the USA, one of the biggest cereal-producing (mainly wheat) areas of the world. *See also* PAMPAS; STEPPES; VELD.

precipitation the depositing of moisture from the ATMOSPHERE onto the Earth's surface. Precipitation may be DEW, drizzle, RAIN, hail, sleet or SNOW. It occurs when the air is cooled and can no longer carry the same volume of water vapour.

prevailing wind the wind direction most often recorded at the location under consideration.

prime meridian *see* MERIDIAN.

progradation the situation that occurs when the BEACH BUDGET is upset and more material is deposited than is removed. The net result is that the beach and shoreline advance towards the sea. Progradation may occur naturally as the result of LONGSHORE DRIFT or estuarine deposits, or it may occur after a sea wall has been built nearby. *See* RETROGRADATION.

pumped storage scheme a system frequently used in HYDROELECTRICITY generation. Water that is used during the day to turn the generators is pumped back up to the reservoir during the night when demand for electricity is low. It helps to conserve water supplies and to ensure that, even in drought conditions, there will be a sufficient head of water to allow the generators to function.

Q

quadrat an area in which a sample is surveyed. For convenience, quadrats are usually small, square areas. Some attribute, e.g. vegetation, in one quadrat can be compared with the same attribute in another quadrat when building up an overall picture of the area.

quagmire a bog that is so soft and wet that the surface moves up and down when walked on.

quarry an open pit or rock face from which ROCK, stone, gravel or SAND is taken by excavation.

quartz one of the most widely distributed ROCK-forming minerals, SiO_2. It occurs in all kinds of rocks, and in its various crystalline forms and with certain impurities, it forms semi-precious stones, e.g. amethyst and agate.

quicksand an unstable stretch of super-saturated SAND. Water moves upwards through it so fast that the sand is held in suspension and cannot bear weight. Any heavy object placed upon a quicksand is quickly sucked down through it.

quota a fixed, allowable amount. Quotas are usually imposed in order to try to regulate supplies to markets. Fishing quotas have been imposed on member states of the EEC in an attempt to conserve fish stocks. Milk quotas were imposed by the EEC to curb overproduction.

R

radar (acronym for Radio Detection And Ranging) the use of radio waves to detect the presence and distance of objects and used in navigation of aircraft, ships, missiles and SATELLITES. It is a REMOTE SENSING system that is of great value in geological, terrain and land use studies.

radiosonde an instrument that is used in meteorology and is carried through successive atmospheric levels by a balloon. The apparatus measures temperature, HUMIDITY and pressure, and the results are transmitted to a radio receiver.

rain one form of PRECIPITATION in which drops of water condense from the water vapour in the atmosphere to form rain drops. Other types of precipitation, all water in some liquid or solid form, include SNOW, hail, sleet, drizzle and also DEW. Snow forms below 0°C and, depending on the temperature, occurs in different shapes. When the temperature is well below freezing, it forms ice *spicules*, which are small and needle-like. Nearer to 0°C, the characteristic snowflakes grow, but at extremely low temperatures snow becomes powdery. Because snow can vary in its form and accumulation, accurate measurement of falls is difficult, but 25 millimetres of water will be produced by about 300 millimetres of newly fallen snow.

Hail is a small pellet of frozen water that forms by rain drops being taken higher into colder parts of the atmosphere. As it then falls, the hailstone grows by adding layers of ice, as a result of condensation of moisture upon the cold nucleus. *Dew* is the condensation of water vapour in the air caused by a cooling of the air.

rain shadow the production of dry, or even desert, conditions on the landward side of MOUNTAINS because most of the moisture from the winds blowing off the ocean or sea falls on the slopes facing the ocean. This occurs in the USA, where the desert areas of Nevada and eastern California contrast with the wet, western side of the Coast Range and Sierra Nevada.

raised beach a beach that is now above the level of the shoreline. This may be because of earth movements or a fall in sea level.

rapids a part of a RIVER where the GRADIENT steepens and flow is fast and turbulent. There may be very small waterfalls, exposed BOULDERS and much broken water. *See* CATARACT.

ravine a deep, narrow VALLEY, bigger than a GULLY, the sides of which slope more than in a GORGE.

reach (1) a narrow inlet of the sea. (2) a straight stretch of a RIVER between bends or locks.

reef (1) any line of rocks in the sea that is exposed at low water but submerged at high water. (2) a vein bearing gold or other precious metal, especially in South Africa. (3) a CORAL REEF that may form an ATOLL.

relative humidity the amount of water vapour in the air compared to the maximum amount of water vapour which that volume of air at the same temperature could carry. It is expressed as a percentage. Saturated air has a relative humidity of 100 per cent. (*See also* ABSOLUTE HUMIDITY).

relief the shape of the land surface of the Earth. *Low relief* means that there are no great differences of height in the area being considered. *High relief* indicates mountainous terrain. A *relief map* shows all the different heights and slopes by means of CONTOUR lines, HACHURES or shading.

relief rainfall *or* **orographic rainfall** when moisture-laden clouds are forced to rise in order to cross mountains, the air cools, water vapour condenses and falls as relief rain. It tends to fall on the seaward side of a mountain range. *See* RAIN SHADOW.

remote sensing the collection of a variety of information without contact with the object of study. This includes aerial photography from both aircraft and SATELLITES, and the use of infrared, ultraviolet and microwave radiation emitted from the object, e.g. an individual site, part of a town, or crop and forest patterns. Another type of remote sensing involves the production of an impulse of light, or RADAR, which is reflected by the object and the image is then captured on film or tape.

Using these various techniques, large areas of the ground can be studied and surprisingly sharp pictures obtained that can be used in many ways. Remote sensing is used in agriculture and forestry, civil engineering, geology, geography and archaeology, amongst others. In addition, it is possible to create pictures with a remarkable amount of detail, which would otherwise take a very long time to collect.

renewable resource a resource that keeps replacing itself, e.g. renewable resources such as the wind, tide, or heat from the Sun could be used to generate electricity for ever, whereas coal and oil, FOSSIL FUELS and therefore non-renewable sources of power, will eventually be used up. Properly managed fish stocks are a renewable resource as sufficient numbers are left to allow them to breed. Similarly, properly managed forestry is a renewable resource as young trees are planted to replace those that are felled.

representative fraction (RF) the ratio of one unit of length on a map to the distance represented by that unit on the ground. The most common ORDNANCE SURVEY maps have an RF of

1:50 000, i.e. 1 centimetre to 0.5 kilometre. This replaced the maps with 1 inch to 1 mile, an RF of 1:63,360.

reserves resources that can be developed economically. This depends on the demand, the costs involved and the state of technology. They may be exploited immediately or held for future exploitation. For example, the technology has only just become available to exploit the oil reserves in the Atlantic.

retrogradation when a SHORELINE is eroded by the sea and retreats inland. This is more likely to occur where the land consists of soft material rather than resistant ROCK. A most notable example of retrogradation is at Holderness, where the shoreline retreated at more than two metres per year during the 1980s. The north Yorkshire coast is also being removed at an alarming rate.

RF abbreviation of REPRESENTATIVE FRACTION.

ria a coastal RIVER VALLEY or ESTUARY that has been flooded as a result of a rise in sea level. Most rias are found on DISCORDANT coasts. They are V-shaped, deepen as they approach the sea and contain many spurs. The mouth of the river Fal in south-west England is a good example.

ribbon development *or* **linear development** the development, often only one plot deep, running along a main route away from a city centre.

Richter scale the scale, devised by the American seismologist Charles Francis Richter (1900–1985), used to measure the intensity of EARTHQUAKES. It uses the amplitude of seismic waves, which depends on the depth of the earthquake focus. Recording stations register the waves, and for a shallow earthquake the magnitude is given by:

$$M = \log (a/t) + 1.66 \log\Delta + 3.3$$

where a is the maximum amplitude, t the period (the time between a repeat of the same wave form) and Δ is the angular distance between the focus and the station. A slightly modified version is used for deeper earthquakes. Earlier systems of measuring intensity, e.g. that devised by Giuseppe Mercalli (1850–1914), the Italian seismologist, relied more on the effects seen or felt by observers when the seismic waves reached them. Below, in brief, is the arbitrary scale from 1 to 12:

1	*Instrumental*	detected only by seismographs.
2	*Feeble*	noticed by sensitive people.
3	*Slight*	similar to a passing lorry.
4	*Moderate*	rocking of loose objects.
5	*Rather strong*	felt generally.
6	*Strong*	trees sway; loose objects fall.
7	*Very strong*	walls crack.
8	*Destructive*	chimneys fall; masonry cracked.
9	*Ruinous*	collapse of houses where ground starts to crack.
10	*Disastrous*	buildings destroyed; ground badly cracked.
11	*Very disastrous*	bridges and most buildings destroyed; landslides.
12	*Catastrophic*	total destruction; ground moves in waves.

ridge of high pressure an elongated area of high pressure between two DEPRESSIONS. It brings a short spell of fine weather.

rift valley a VALLEY between two parallel FAULTS, formed when the rock between the faults sank. There are many examples, from the Red Sea to parts of the Rhine valley.

rill a small, narrow, steep-sided stream that is often dry. However, after storms they can carry enough water to cause soil EROSION. It has been suggested that, as parallel rills become bigger, they join together along their length to form a GULLEY.

rip a turbulent area of sea water where strong tidal CURRENTS meet, or where a tidal current meets wind-driven waves head on.

rip current a strong, swift CURRENT. When waves tend to pile up water against a coast, a current runs parallel to the coast until it meets an obstruction such as a HEADLAND or SPIT. The current is deflected seaward and becomes a rip current.

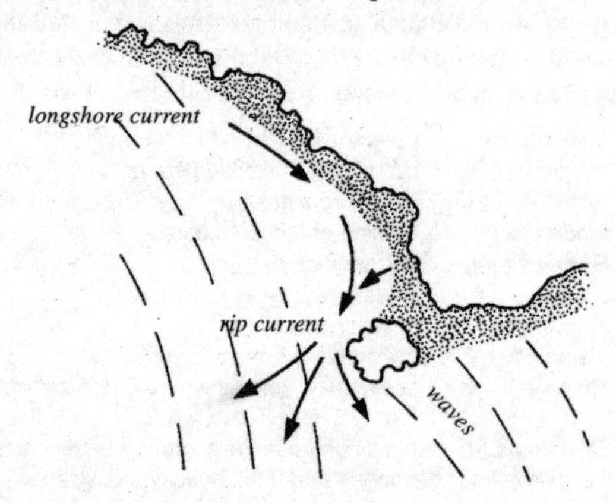

river capture literally, the capture or diversion of one river into an adjacent river, often at a sharp change of stream direction. One river enlarges its channel upstream where it meets the headwaters of the other river. If the latter is a less energetic system the waters may then be diverted.

rivers streams of water that flow into the sea or, in some cases, into LAKES or swamps. Rivers form part of the cyclical nature of water, comprising water falling from the ATMOSPHERE as some form of PRECIPITATION, and being partly fed by groundwaters or run-off from the melting of GLACIERS (both of which in any case are derived from atmospheric water).

Rivers develop their own immediate scenery and a river VALLEY will owe its form to the original slope of the land, the underlying ROCKS and the CLIMATE. A river with its tributaries is called a *river system*, and the area from which its water is derived is the *drainage basin*. As rivers grow in size and velocity, rock and soil debris washed into them is carried downstream, eroding the river bed and sides as it goes. As a river continues to flow and carry debris, depositing much material in times of FLOOD, it widens its valley floor, forming a FLOOD PLAIN. As it does, the river swings from side to side, forming wide loops called MEANDERS. Eventually, as meanders develop into ever more contorted loops, a narrow neck of land may be left that is eventually breached. Thus the river alters and shortens its course, leaving a horseshoe-shaped remnant, or OXBOW LAKE.

Some of the World's longest rivers

		km
Nile	Africa	6695
Amazon	South America	6516
Yangtze	Asia	6380
Mississippi	North America	6019
Murray-Darling	longest in Australia	3750
Volga	longest in Europe	3686

(the longest in the UK is the Severn, 350 km)

river terrace a raised flat area running along and parallel to VALLEY walls. It is part of a former FLOOD PLAIN, left behind as the RIVER cut down through the valley floor. A series of terraces indicates periods of aggradation followed by periods when the river eroded its own river bed.

road *or* **roadstead** an anchorage near an estuary or harbour where there is some shelter in stormy conditions, e.g. Yarmouth Roads is the area between the shore and the Scrobie Sand, an offshore sandbank.

Roaring Forties the westerly gales that are a feature of the

OCEANS 40°–50° south. At these latitudes there are no land masses either to break up heavy seas or to interrupt the wind, with the result that stationary pressure systems do not develop. DEPRESSIONS are frequent.

rocks aggregates of MINERALS or organic matter, which can be divided into three types, based on the way they are formed: IGNEOUS, SEDIMENTARY and METAMORPHIC.

roller a colloquial term for a large OCEAN wave that breaks on exposed coastlines. Good rollers are important in the sport of surfing. Some of the best are to be found round the Pacific islands and along the coast of California.

Rossby waves the wave patterns formed by the wind in the mid to upper TROPOSPHERE as they travel at the edge of a polar front (*see* POLAR AIR MASS). There are usually four, five or six of these waves, with a wavelength of about 2000 kilometres (1250 miles), travelling round the high and mid-latitudes of

Rossby waves

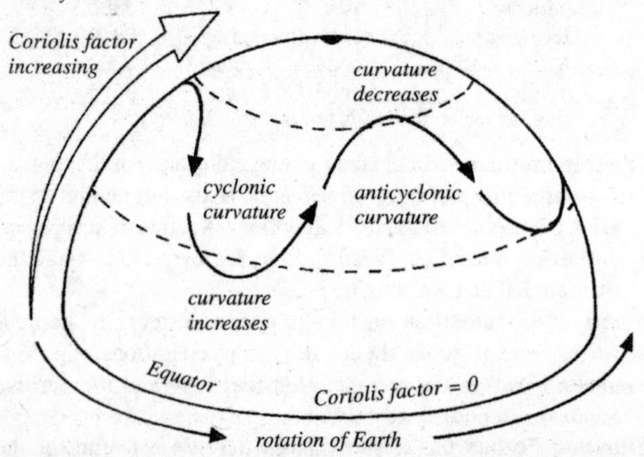

the EARTH. They seem to be caused by the interaction between the ROTATION of the Earth, and winds caused by differential heating of the ATMOSPHERE.

The cycle starts as a strong west-to-east airflow with little mixing of polar and TROPICAL AIR MASSES. As the waves develop, so does a meandering pattern with north-to-south and south-to-north air flows.

In the northern HEMISPHERE, troughs of low pressure develop on the left of a south-to-north flow as the polar air mass moves southwards. At the same time, a RIDGE OF HIGH PRESSURE develops on the left of a north-to-south flow and a tropical air mass moves northwards. The system then breaks down, transferring 'patches' of polar and tropical air across the polar front, and the cycle starts again with strong westerlies.

rotation of the Earth the motion of the Earth from west to east as it rotates on its AXIS. The velocity of rotation is greatest at the EQUATOR—1690 kilometres (1056 miles) per hour—and decreases to zero at the poles. It is responsible for creating the predominantly west-to-east passage of weather systems.

runoff the water that leaves a drainage area. It is usually estimated as the rainfall minus any water lost by evaporation. Runoff occurs after heavy rain when the ground is saturated and can absorb no more water.

rural pertaining to the countryside.

S

sabkha a flat, coastal belt situated between desert DUNES on the landward side and a LAGOON and the sea on the other side. It is a site for the formation of evaporite deposits, notably carbonates and sulphates. It is named after the Trucial Coast in the Persian Gulf, *sabkha* being the Arabic word for 'salt flat'.

Sahel the narrow belt of semi-desert that stretches across Africa from Senegal to the Red Sea. It lies south of the Sahara and north of the grasslands. Annual rainfall is from 150 millimetres to 500 millimetres (5 to 20 inches), but droughts, with their attendant crop failures and famines, are frequent.

salinization the process whereby soils tend to become salty when there is a high rate of EVAPORATION from the surface. Evaporation tends to draw water, along with any salts in solution, up to the surface of the soil. The salts are left when the water evaporates. Salinization also tends to occur on irrigated land in hot climates.

saltation the process of moving particles by water or wind that are too heavy to remain in suspension. The particles, usually of sand, tend to bounce along the surface as they are blown by the wind. Particles of grit, too heavy to be carried in the CURRENT of a RIVER, may nevertheless be light enough to be disturbed by the current and bounced along the river bed. In both cases, as a particle lands, the impact may be enough to dislodge other particles.

salt dome a plug-shaped body, either circular or elongated, that is formed by the upwards movement of lighter evaporitic sediments into the overlying, denser rocks. The evaporite is usually halite. Salt domes may be 1 or 2 kilometres (1100–2190 yards) in diameter, but extend to great depths.

salt flat the flat, dried-out bed of a former SALT LAKE.

salt lake a LAKE containing water that has become very salty because of a high rate of EVAPORATION. It is usually much saltier than sea water. For example, the Dead Sea is approximately one part salt to four parts water while sea water is about one part salt to 32 parts water.

salt marsh a coastal MARSH that forms on mud flats. It may be tidal and flooded twice a day by the sea. The vegetation is adapted to survive regular sea-water flooding, freshwater flooding and exposure to the air at low tide.

salt pan a very small, shallow SALT LAKE. After EVAPORATION, deposits of salt are left behind that may be exploited commercially. Salt pans are often artificially constructed in flat, coastal areas of hot countries. The sea is allowed to flood the pan, which is then shut off while the water evaporates, leaving a layer of salt behind.

sand particles of rock (mainly quartz) having a diameter of 0.06 to 2.0 mm. Sand is required in huge quantities by the building industry, and sand that is especially rich in quartz is the basic material for glass making. Sandy soil is light and easily worked, but allows water to percolate through it very quickly. The sand on beaches is often formed from finely crushed shells. This sand is rich in calcium and is sometimes used to improve the texture of heavy soils.

sand dune a hill that is formed by wind-blown SAND. Once a mound forms, it tends to grow because the friction of the sandy surface slows down the wind, which then cannot carry so much material. The excess is deposited on the dune. *Parabolic dunes* curve around a patch of vegetation, with the bend pointing downwind. *Seif dunes* are longitudinal with a sharp ridge. They form when two prevailing winds alternate, possibly seasonally. *Star dunes* are huge pyramidal dunes with radiating ridges that form where wind direction is very changeable. *Barchans*, small curved dunes the horns of which point downwind, form where there is not much loose sand. *Wind shadow dunes* form in the LEE of hills, while *lunettes* form in the lee of DEFLATION HOLLOWS. *Stabilized dunes* are coastal dunes that are covered in vegetation (*see* MARRAM GRASS).

sandstone a SEDIMENTARY ROCK comprising SAND grains with sizes between 0.06 and 1 millimetre, and a variety of cements and other minerals. Calcite is a common cement, and silica (as quartz) cements sands to produce a hard sandstone often referred to as an *orthoquartzite* (to differentiate from the METAMORPHIC ROCK, quartzite). Brown and red sandstones usually have an iron-rich cement such as limonite or haematite. Other minerals that may be present include feldspar and mica.

saprolite weathered rock in its place of origin. Formation depends on a stable combination of high temperature and rainfall within a gentle landform, thus minimizing the removal of the weathered rock. Saprolites are thus best formed under tropical conditions, and large thicknesses may be generated.

satellite an object that revolves in orbit around another, larger object. The first artificial satellite was launched in 1957 to orbit the EARTH. Artificial satellites are packed with scientific equipment that collects information about the planet they are orbiting. That information is then transmitted to Earth, where images are formed using REMOTE SENSING techniques. The information obtained by satellites about CLOUD formations has done much to improve weather forecasting. The MOON is a satellite of the Earth.

savanna *or* **savannah** is similar to grassland but with scattered trees and is found extensively in South America, southern Africa and parts of Australia. There are usually well-defined seasons: cool and dry, hot and dry, followed by warm and wet, and during the latter there is a rich growth of grasses and small plants. Although savanna soils may be fertile, they are highly porous and water therefore drains away rapidly.

scarp *see* ESCARPMENT.

scarp-and-vale terrain the undulating pattern formed by a series of CUESTAS and VALLEYS where the SEDIMENTARY ROCK is tilted gently. The softer rocks, e.g. CLAY, are eroded to form the valleys, while the harder limestones and SANDSTONES form the cuestas.

science park a relatively recent industrial development in which sites are created to house and cater for high technology ventures and companies undertaking research and development. Science parks are often linked to, or developed by, universities with a traditionally strong base in science and technology.

scirocco *see* SIROCCO.

Scotch mist a fine drizzle falling from a thick, low CLOUD base. It is experienced on hills and MOUNTAINS when saturated, stable AIR MASSES are forced to rise, so causing CONDENSATION to occur.

scour the ability of a CURRENT of water to cause EROSION. A tidal current may cause scouring in an ESTUARY or in STRAITS. A RIVER current may cause scouring on the outside of a bend. In both cases, the material removed will be dumped where the current slackens, a process known as *scour-and-fill.*

scree the accumulation of mainly coarse, angular rock debris at the foot of cliffs inland. The debris is produced by the weathering and gradual disintegration of the upper slopes and cliffs through the agencies of frost and water.

scrub the vegetation that grows on poor soil in semi-desert conditions. It is coarse, sparse, gnarled or stunted and adapted to survive periods of DROUGHT.

sea, law of the an agreement between most maritime nations for administering the seas. It recognizes and defines internal waters, TERRITORIAL SEAS, the CONTIGUOUS ZONE, the CONTINENTAL SHELF and the high seas. It also recognizes exclusive fishing and economic zones of up to 320 kilometres (200 miles) from a nation's coastline.

sea breeze *see* LAND AND SEA BREEZES.

SEASAT an artificial SATELLITE, launched in 1978 with the object of collecting information on the OCEANS using REMOTE SENSING techniques. It had transmitted a vast amount of data when it fell from orbit a few months after it was launched.

sedimentary rock one of the three main rock types. They can be divided into *clastic rocks*, which are made of fragments, *organic rocks* and *chemical rocks*.

Typical sedimentary rocks

Clastic	*Organic*	*Chemical*
sandstone	LIMESTONE	limestone
shale	ironstone	dolomite
mudstone	chert	flint
conglomerate	COALS	gypsum and other EVAPORITES
limestone		

The clastic rocks are further divided on grain size into *coarse* (or *rudaceous*, grains of 1–2 millimetres), *medium* (or *arenaceous*, e.g. sandstone) and *fine* (or *argillaceous*, up to 0.06 millimetre). When the grains comprising clastic rocks are deposited (usually in water) compaction of the soft sediment and subsequent *lithification* (i.e. turning into rock) produce the layered effect, or *bedding*, that is often visible in cliffs and outcrops in rivers. It is also common for original features to be preserved, e.g. ripples, small or large dune structures, which in an exposed rock face appear as inclined beds called *current bedding*. *Graded bedding* shows a gradual change in grain size from the base, where it is coarse, to the top of a bed, where it is fine, and this is because of the settling of material onto the sea floor from a current caused by some earth movement.

Many sedimentary rocks, particularly shale, limestone and finer sandstone, contain FOSSILS of animals and plants from millions of years ago, and with the original features mentioned above, these are useful in working out the sequence of events in an area where the rocks have been strongly folded.

seismograph in the study of EARTHQUAKES (*seismology*), seismographs are used to record the shock waves (*seismic waves*) as they spread out from the source. The seismograph has some means of conducting the ground vibrations through a device to turn movement into a signal that can be recorded. There are numerous seismic stations around the world that record ground movements, each containing several seismographs with numerous *seismometers* (the actual detector linked to a seismograph).

seismic survey a technique used to investigate the nature of underground rocks, especially those that may contain valuable ores, minerals or oil. A SEISMOGRAPH is used to measure the time required for a pulse, emitted on the surface of the ground, to be reflected back from a discontinuity in the underlying rocks. A series of readings enables a picture of the structure of the ground to be built up.

selva thick TROPICAL RAIN FOREST, e.g. in the Amazon basin.

semi-diurnal tide high tide and low tide occur twice in each lunar day, i.e. high tide occurs 12 hours 25 minutes after the previous high tide. Although tides are governed by the position of the Moon, the configuration of the land determines how often the tide turns.

sensor a device that can detect a force or a part of the electromagnetic spectrum. Sensors are an essential part of REMOTE SENSING technology. Examples are RADAR and cameras used in aerial photography.

service industry those economic activities concerned with the distribution and consumption of goods and services, not with their manufacture or production. For example, manufacturing industry produces washing machines but the service industry is responsible for selling them. Education, health and insurance are further examples of service industries.

set-aside (1) a government grant from the government to a farmer in an EC country so that the farmer will take a field or fields out of agriculture for an agreed period of time. It was hoped that surpluses would be cut down, but in practice many farmers increase production on other fields. (2) the land that has been taken out of production in exchange for a set-aside grant.

shade temperature the temperature of the air recorded inside a STEVENSON SCREEN, away from direct sunlight or other radiation. All temperatures quoted in meteorology are shade temperatures unless specified otherwise.

shale a SEDIMENTARY ROCK that is fine-grained (composed of silt, mud and clay-sized particles). Shales are fissile because of the alignment of clay and similar minerals with their flat surfaces parallel to the planar fabric.

shanty town a haphazard collection of shacks, usually erected without permission on the outskirts of Third World cities. They grow as demand for cheap housing far outstrips supply. Roads are little more than dust tracks, and services such as education or sanitation and medical facilities are inadequate. Electricity, gas or piped water may well not be available.

shield a part of the continental crust that is stable and has been unaffected by MOUNTAIN building for a long time. The ROCK in a shield is usually more than 570 million years old.

shoal a BANK of mud or SAND that lies just under the surface of the sea and is a danger to shipping.

shoreline the contact line between land and the water in a LAKE, sea or OCEAN. It is not equivalent to 'coastline', as the latter is not used of lake shores.

shore platform the gentle slope extending from the base of a cliff towards the sea. It is subject to both wind and wave EROSION. It is argued that shore platforms wider than 800 metres must have been formed, in part at least, by a fall in sea level.

Siberian high a persistent ANTICYCLONE that forms over central and northern Asia during the winter. It is one of the sources of polar continental AIR MASSES. Extensive snow cover and the huge land mass ensure that winter temperatures are extremely low.

sierra the Spanish term for a range of MOUNTAINS with jagged peaks. It is common in the names of jagged mountain ranges where there is or was Spanish influence, e.g. the Sierra Nevada in eastern California.

silt a layer of very fine particles, coarser than CLAY but finer than SAND. Particles of silt have a diameter of between 0.002 and 0.06 millimetre. Silt is often carried and deposited by RIVERS.

sink-hole a funnel-shaped depression in CHALK or LIMESTONE country. It was probably caused by the subterranean collapse of a cave. Although sink-holes are usually dry, a few have a stream draining into them.

sirocco *or* **scirocco** a hot, dust-laden wind blowing northwards from the Sahara in advance of DEPRESSIONS that move eastwards across the Mediterranean. Initially, it is a dry wind blowing into Spain but, as the depression pushes it eastwards, it picks up moisture from the Mediterranean Sea. By the time it reaches Italy it has become humid and brings oppressive conditions.

Site of Special Scientific Interest (SSSI) an area in Britain that is of particular importance because of its landscape, vegetation, fauna or geology. Development within an SSSI is carefully controlled. There are many SSSIs of varying sizes throughout the country.

skerry a small, offshore, rocky islet of northern Europe. The word is used in place names in Scotland and Scandinavia.

slack (1) the period of still water (*slack tide*) between ebb and flow CURRENTS. (2) the portion of a RIVER current that moves more slowly, often on the inside of a bend (*slack water*). (3) the hollow between coastal sand DUNES.

smog a word formed from 'smoke' and 'fog', dating from the early years of the 20th century. Smog was formed by the smoke from domestic chimneys along with smoke, sulphur dioxide and other pollutants from industrial emissions. If an *inversion layer* (a layer of cold air overlying a layer of warm air) formed over a city or industrial area, pollutants were unable to escape upwards into the atmosphere. They formed a thick, yellowish acrid smog that, each winter, caused many deaths among people with respiratory problems. Eventually, in the middle of the 20th century, 'clean air' Acts of Parliament were

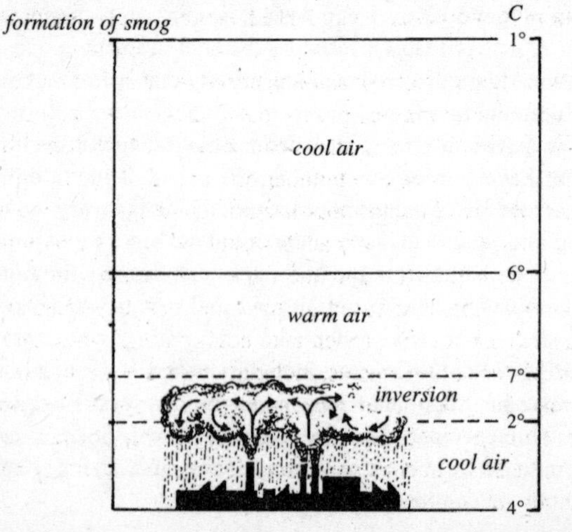

formation of smog

passed to combat the problem, smokeless fuel was introduced and pea-soup smog became a thing of the past.

The word smog is now often used to describe the petrochemical fog that forms over cities, mainly as a result of vehicle exhaust fumes.

snow white flakes of frozen water vapour. At very low temperatures, snow is powdery and dry, the separate flakes do not stick together and snow is relatively light to shift, e.g. in clearing a drift. At temperatures nearer 0°C, snow is heavy and wet. The flakes, which take various geometric patterns, stick together. If wet snow freezes onto power lines, the cables are likely to sag under the weight and may bring down the poles carrying them with resultant loss of the electricity supply. The type of snow is important to the skiing industry, dry, powdery snow giving the best conditions.

social capital those assets that belong to society as a whole, rather than to individuals, e.g. NHS hospitals, schools, roads, etc.

soil the thin layer of uncompacted material comprising organic matter and minute mineral grains that overlies ROCK and provides the means by which plants can grow. Soil is formed by the breakdown of rock in a number of ways. Rock is initially fractured and broken up by WEATHERING, which is the action of water, ice and wind, and any acids dissolved in water moving over, or percolating through the rock. This allows entry to various organisms that speed up the breakdown process and mosses, lichens and fungi then take hold. After a while there forms a mixture of organisms, including bacteria, decayed organic material, weathered rock and HUMUS, which is called TOPSOIL. Humus is decomposing (breaking down) organic material produced from dead organisms, leaves and other organic material by the action of bacteria and fungi.

The texture of the soil affects its ability to support plants, and the most fertile soils are *loams*, which contain mixtures of sand, silt and clay with organic material. This ensures there is sufficient water and minerals (which 'stick' to the finer particles), while the coarser sand grains provide air spaces, vital to roots. In addition to plant roots, soil contains an enormous number of organisms, including fungi, algae, insects, earthworms, nematodes (roundworms) and several billion bacteria. Earthworms are useful in that they aerate the soil, and the bacteria alter the mineral composition of the soil.

The parent rock is the primary factor in determining the nature of a soil. While sand, silt and clay produce a loam, sand alone is too porous and clay too compacted and impervious (doesn't allow water through). A clay soil can be improved by adding lime, hence a *marl* (a lime rich clay) forms a good soil. Limestone itself does not produce a soil. The rate of breakdown of the rock is also important. Granites decompose slowly, but basaltic rocks are the opposite and therefore yield their soil components quickly. This is seen particularly in volcanic areas where lava flows and volcanic ash quickly lead to very productive soils.

It can take hundreds of years for soils to become fertile, but to be productive agriculturally, the soil has to be cared for, with IRRIGATION, fertilization and prevention of EROSION all being important factors. This is apparent when you consider that the soil provides approximately 18 kilograms of nitrogen, 4 kilograms of potassium and 3.5 kilograms of phosphorus to grow one ton of wheat grain.

solar pertaining to the Sun. A *solar day*, 24 hours, is the time taken for the same MERIDIAN on Earth to make two successive crossings under the Sun, i.e. noon to noon at the same place. The *solar year* is the time taken by the Earth to make one complete orbit round the Sun. It is 365 days, 5 hours, 48 minutes and 45.51 seconds long.

solar energy energy of any form that is derived from the Sun. *Solar cells* convert the Sun's radiation to electricity using thermoelectric or photovoltaic devices. *Solar panels* contain a fluid, such as water, that is heated and then circulated throughout a building to provide heating. Although the Sun, in theory, is an inexhaustible supply of energy, the technology does not yet exist to convert its radiation to electricity on a large scale, e.g. to supply the needs of industry. Plants use solar energy to convert carbon dioxide and water into carbohydrates.

solar radiation the heat, light, X-rays, gamma rays and ultraviolet rays emitted by the Sun. The heat and light radiation are responsible for the Earth's CLIMATE. More radiation is received at the EQUATOR than at the poles. Much of the ultraviolet radiation is absorbed by the OZONE LAYER.

solstice the time at which the Sun reaches its most extreme position north or south of the EQUATOR. There are two such instants in the year.

solar wind the term for the stream of charged, high-energy particles (primarily electrons, protons and alpha particles) emitted by the Sun. The particles travel at hundreds of kilometres per second, and the wind is greatest during flare and sunspot activity. Around the Earth, particles have velocities of 300–500 kilometres^{-1}, and some become trapped in the magnetic field to form the Van Allen radiation belt. However, some reach the upper ATMOSPHERE and move to the poles, producing the auroral displays (*see* AURORA).

solifluction the downhill movement of water-saturated regolith (the layer of unconsolidated and weathered material that lies over solid ROCK. It may comprise rock fragments, mineral grains and soil components, and in the humid tropics can reach enormous thicknesses—commonly tens of metres). It was first described in periglacial areas (i.e. those next to a GLACIER or ICE SHEET) but now applies to all environments. It is particularly prevalent in the humid conditions of the tropics.

sonde a REMOTE SENSING device used at relatively low levels to gain information about the ATMOSPHERE.

sonic mapping the use of an echo sounder to produce a map showing the depths of the sea. The echo sounder produces a high frequency signal that travels to the seabed and is bounced back to the ship. The depth of the sea at that point can be calculated from the time between emitting the pulse and receiving it back.

South Pole the southern end of the Earth's AXIS.

spate a sudden flood and increase in speed in the CURRENT of a RIVER. Spates occur after heavy rain or a sudden thaw near the headwaters of a river. They can cause FLASH FLOODS, which may do much damage downstream. Spates are usually short-lived.

spa town a town that has grown up around mineral SPRINGS. These springs are believed to have therapeutic value and, in Victorian times, attracted numbers of visitors who came to drink the water or to bathe in the specially constructed indoor pools. Bath, Harrogate and Strathpeffer are examples, although nowadays people neither drink the water nor bathe in it as regularly. However, on the Continent, people continue to be convinced of the therapeutic value of mineral springs and spa towns are always popular holiday resorts, Baden-Baden being one of the best known.

specular reflector in REMOTE SENSING, a surface that reflects electromagnetic radiation without scattering or diffusing it, e.g. a still water surface.

spit a narrow, elongated ridge of SAND and shingle sticking out into the sea. Spits grow as the result of LONGSHORE DRIFT, usually near the mouth of an ESTUARY where the coastline changes direction.

spring (1) a natural flow of water from the ground. It marks the point where the top of the WATER TABLE meets the ground. Spring water is usually very pure but may contain dissolved mineral salts. In recent years, a considerable industry has sprung up to bottle spring water for sale. *See* HOT SPRINGS; MINERAL SPRINGS. (2) the season between winter and summer.

spring tide a tide with a maximum range between high and low water. Spring tides occur twice each month when the Sun, Moon and Earth are in a straight line. The biggest effect is seen at the time of the new moon when the Sun and Moon are on the same side of the Earth.

squall a sudden, violent gust of wind during a storm. Wind speed must rise by at least 8 metres (26 feet) per second and reach 11 metres (36 feet) per second before dying away. The gust should last for at least one minute, although the squall will last much longer. Squalls are accompanied by heavy rain.

SSSI abbreviation of SITE OF SPECIAL SCIENTIFIC INTEREST.

stack a high, steep ROCK pillar close to a cliff but separated from it by the sea. A stack is formed when the ARCH joining it to a cliff collapses. The rock in a stack is more resistant to EROSION by the waves than the material that once joined it to the land. The most spectacular example in Britain is the Old Man of Hoy in Orkney.

stalactites and stalagmites in areas of LIMESTONE where caves form and streams trickle through the rocks and caves, calcium-rich waters tend to drop from cave roofs. As there is a little evaporation from these drops of water, some of the dissolved calcium is deposited as calcite (calcium carbonate, $CaCO_3$). This deposit builds up very slowly into a *stalactite* projecting down from the roof. If water continues to drop to the floor, a complementary upward growth develops into a *stalagmite*—and often the two meet to form a column, or pillar.

Many limestone caves exhibit spectacular developments, e.g. White Scar at Ingleton in Yorkshire; La Cave in the Dordogne; Wookey Hole in the Mendips, and many other places.

stand a grouping of trees of the same type in a forest. If trees are planted in stands then felling is an easier operation.

standard atmosphere an accepted condition of the atmosphere for use in calibrating instruments and measuring altitudes. The most commonly accepted standard atmosphere is based on a surface temperature of 15°C and surface pressure of 1013.25 millibars or 760 millimetres of mercury.

standard time the time, based on the Sun's position at noon, that is used for timekeeping within a country. The world is divided into a number of time zones, each differing from the Greenwich MERIDIAN by a multiple of 15° or an exact number of hours. To find standard time for a country, the time is calculated at a meridian that is roughly central to the country, and this is the time used within that country. Very large countries, such as the USA, have several time zones.

staple the most important item in an economy. This may be the local food crop, e.g. rice in Asia, dates in north Africa, potatoes in northern Europe, or it may be an export, e.g. bananas from the West Indies or electronic goods from Japan.

steppes the mid-latitude grasslands that stretch from central Europe towards Siberia. The steppes were flat or rolling, treeless plains, now used for agriculture. *See* PAMPAS; PRAIRIES; VELD.

Stevenson screen a box designed by Robert Louis Stevenson's father to protect meteorological instruments from conditions that might lead to inaccurate readings. It is a white-painted wooden box, raised on legs about one metre from the ground. The sides are louvred to enable air to circulate freely around the instruments without them being exposed to direct sunlight. The opening side is to the north in the northern hemisphere. It normally contains a variety of different thermometers and a hydrograph.

storm surge if there is an area of extremely low pressure over the sea, sea levels tend to rise. If the winds associated with a DEPRESSION blow towards a coastline, the net result will be a build-up of water on that coast. If this occurs during a period of high SPRING TIDES, the sea level is likely to be exceptionally high and may cause extensive flooding and damage. As the result of such conditions, a particularly disastrous storm surge occurred in the southern part of the North Sea in 1953. In eastern England and the Netherlands flooding was extensive. Much damage was done and many people were drowned.

strait a narrow stretch of sea linking two larger seas, e.g. the Straits of Dover between the North Sea and the English Channel.

strath the Scottish term for a flat-floored RIVER VALLEY. It is broader than a GLEN.

stratosphere one of the layers of the atmosphere, lying above the TROPOSPHERE. It lies at a height of between 10 and 50 kilometres (6–31 miles) and shows an increase in temperature from bottom to top, where it is 0°C. (The average temperature above is –60°C). A very large part of the ozone (*see* OZONE LAYER) in the atmosphere is found in the stratosphere, and the absorption of ultraviolet radiation contributes to the higher temperature in the upper reaches. This inversion of temperature creates a stability that limits the vertical extent of CLOUD and produces the sideways extension of a cumulonimbus cloud into the characteristic shape.

stratum (*plural* **strata**) a layer or bed of ROCK that has no limit on its thickness.

stratus a spread-out CLOUD form with an even base and generally grey appearance, through which the Sun may be seen providing the cloud is not too dense. It occasionally forms ragged patches.

sub-Arctic near to, but outside, the ARCTIC CIRCLE.

sublittoral zone the zone of the sea that stretches from the lowest mark of ordinary TIDES to the edge of the CONTINENTAL SHELF.

submerged coast a coastline formed by a rise in sea level or by the land sinking. There will be DROWNED VALLEYS, RIAS, FIARDS or FJORDS on a submerged coast.

subsistence economy an undeveloped economy in which people barter goods rather than buy and sell goods using cash. People live, literally, hand to mouth with little, if any, savings.

subsistence farming a form of farming where all the produce goes to support the household. There is none left over for sale.

subsoil the layer of weathered soil that is immediately above the BEDROCK and below the TOPSOIL.

subtropical the term used to describe those zones near to, but outside the tropics. They stretch from the TROPICS of Cancer and Capricorn to 40° north or south of the EQUATOR.

subtropical high an area of almost constant high pressure about 30° north and south of the EQUATOR. It is formed as warm equatorial air rises, spreads out and then sinks. As it falls it is warmed adiabatically (*see* ADIABATIC TEMPERATURE CHANGE), so rainfall is unlikely. The subtropical ANTICYCLONES formed in this way affect both the TRADE WINDS and the westerly winds and may control the surface winds of the Earth.

sudd a term used in northeast Africa to describe a mass of floating vegetation that has broken away from nearby SWAMPS. It regularly blocks the main channel of the White Nile, causing the RIVER to branch, to form lakes and marshes, and is a danger to navigation.

sunshine recorder *or* **Campbell-Stokes recorder** an apparatus comprising a glass sphere that focuses the Sun's rays on card marked with hours. The focusing creates sufficient heat to burn a track on the card, thus recording the duration of the sunshine.

supply the availability of goods or services. In general, supply will increase as the price received rises. The price will stabilize when the level of supply matches the level of DEMAND.

surf the broken water among BREAKERS near a coast. The surf tends to cover a wider area when breakers approach a sandy beach rather than a shingle beach.

survey any properly organized collection of information about an object, area or population that is being studied. The word is used most often to describe the collection of precise data about the Earth's surface in order to construct a map. Surveys may be undertaken by surveyors on the ground, by REMOTE SENSING or by aerial photography.

swamp a persistently waterlogged area. It differs from a MARSH or BOG in that trees often grow in swamps, e.g. the mangrove swamps of tropical areas.

swash the powerful movement of water up a beach from a breaking wave. It may carry pebbles and cobbles up the beach and plays an important part in beach building as a result of LONGSHORE DRIFT. *See* BACKWASH.

swathe a REMOTE SENSING term to describe the strip of the Earth's surface that is scanned by a sensor in an orbiting SATELLITE.

swell large, smooth waves that travel across the OCEAN. They are set up by wind in one area of the ocean and move quickly across the seas in advance of the wind.

syncline a downfold fold of rocks shaped like a basin, with the younger STRATA uppermost (in the centre).

synoptic chart a weather map showing ATMOSPHERIC PRESSURE, temperature, wind speed and direction, CLOUD cover and PRECIPITATION for a given area at a specific time. From this collection of data, it is possible to predict the movement of FRONTS. Synoptic charts, therefore, are widely used in weather forecasting.

synpotic image a REMOTE SENSING term for the image of a large part of the Earth's surface.

T

taiga the belt of coniferous forest between the temperate grasslands and the TUNDRA. It is more or less the same as the BOREAL FOREST, but the taiga may contain large areas of BOG covered in sphagnum moss.

talik a layer of unfrozen ground between the PERMAFROST and the layer of ground that freezes and thaws. Talik is more common where there are LAKES and RIVERS.

talus *see* SCREE.

tariff a list of fixed charges for a commodity or service. It is the term used for the customs duty to be paid on imported goods. High tariffs may be imposed to discourage imports.

tarn a small mountain LAKE. The term is particularly common in the Lake District.

tectonic concerned with earth movements, as involved in folding and faulting.

temperate climate a moderate climate with no extremes of temperature. It is more likely to be found in MARITIME areas in mid latitudes.

terrestrial (1) pertaining to the planet Earth. (2) being on land rather than on the water.

terrestrial radiation the heat radiated from the Earth. The heat from the Sun warms the surface of the Earth. At night, especially when there is no CLOUD cover, much of this heat is given off, the reason why clear, winter nights are likely to be frosty. About one third of the Sun's radiation received by Earth is radiated back into space.

territorial seas the coastal waters with the sea bed below and the air above them that a state claims as its own. For many years the territorial seas of a nation were taken as three nautical miles from a landward baseline that either followed a smooth coastline or took account of numerous promontories (e.g. in Norway). The Law of the Sea Convention in 1983 proposed an Exclusive Economic Zone of the sea stretching 200 miles (320 kilometres) from the landward baseline. Where there is less than 400 miles of sea between countries, the median line is drawn to mark the extent of their territorial waters. This arrangement applies to the sea between most European countries.

tertiary industry the SERVICE INDUSTRIES, such as education, retailing, hairdressing, etc, that circulate money but do not create new wealth.

thalweg the long profile of a RIVER VALLEY or the line of the deepest part of the stream.

thematic map a map designed to show particular attributes, e.g. population, climate or oil wells, rather than physical features such as mountains or rivers.

theodolite an instrument essential in surveying. It consists of a small telescope that can be rotated in either the vertical or horizontal planes, a spirit level and a compass, all of which are mounted on a tripod.

thermal (1) pertaining to heat. (2) a strong upcurrent of warm air used by gliding birds and glider pilots to gain height.

thermal depression an area of low pressure that occurs in continental interiors. When the Earth's surface gets very hot, warm air rises in CONVECTION currents and ATMOSPHERIC PRESSURE falls at the surface. The resulting thermal depression may be a small, short-lived feature, such as a DUST DEVIL, or it may be a huge feature associated with the MONSOON in India.

thermal erosion (1) the EROSION caused when a frozen RIVER thaws and undercuts its banks. (2) the mass movement of ground when PERMAFROST thaws.

thermal expansion *or* **thermal fracture** when rocks are subject to large daily temperature changes, they crack because the various constituents of the rock expand and contract at different rates.

thermal sensing in REMOTE SENSING, when the SENSOR picks up heat being radiated from a feature, so enabling remote sensing to be carried out in cloudy conditions or in darkness.

thermocline the layer of water in the oceans or deep lakes that separates the warmer surface water from cold, deep water. It is a zone approximately one to three kilometres deep where temperatures fall sharply.

Thornthwaite climate classification (*see also* KÖPPEN CLASSIFICATION) a classification of CLIMATES devised in the early 1930s by C.W. Thornthwaite and based on characteristic vegetations and the associated PRECIPITATION. Two parameters were developed:

P/E and T/E

P/E represents the total monthly precipitation divided by the total monthly evaporation, and T/E is similar, except that it refers to the temperature. Based on the precipitation figures, five humidity provinces were created:

Province	Vegetation	P/E
A	rainforest	>127
B	forest	64–127
C	grassland	32–63
D	steppe	16–31
E	desert	< 16

Each province can be further subdivided by the rainfall pattern, e.g. abundant in all seasons, deficient in winter. The T/E figures provide an additional classification into provinces:

	Province	T/E
A'	tropical	128
B'	mesothermal	64–127
C'	microthermal	32–63
D'	taiga	16–31
E'	tundra	1–15
F'	frost	0

With all this information, Thornthwaite produced a climatic world map of 32 climate types. Each of these areas can be associated with certain soils, geomorphological processes, etc, related to and depending on the climate.

threshold population the minimum population required for the provision of goods or services. In practice, this is difficult to quantify. For example, a small, affluent population may enable a range of restaurants to operate in a given area, whereas a larger but poorer population may not use restaurants at all.

thunder *see* LIGHTNING.

thunderstorm a storm of thunder, LIGHTNING and heavy rain or hail. Sometimes hailstones can be a centimetre or more in diameter. A thunderstorm represents extremely unstable conditions in the ATMOSPHERE caused either by the passage of a cold front or by the rapid rise of a CONVECTION current of warm air from a hot ground surface. For example, thunderstorms occur regularly over Johannesburg during the evenings after hot summer days. If the convection current rises above the CONDENSATION level (the height in the atmosphere at which condensation occurs), CUMULONIMBUS cloud forms. A strong electrical charge builds up in the cloud and, when it discharges to Earth (the lightning), a thunderstorm results. Thunderstorms occur less frequently with distance from the EQUATOR. They are almost unknown in polar latitudes where there is little convection or surface heating.

tides the regular rise and fall of the water levels in the world's oceans and seas that are caused by the gravitational effect of the Moon and Sun. The Moon exerts a stronger pull than the Sun (roughly twice the effect), and variation in tides is caused by the relative positions of the three bodies and the distribution of water on the Earth. When the Sun, Moon and Earth are aligned, the effects are combined and result in a maximum, the high *spring tide* (when the Moon is new or full). Conversely, when the Sun is at right angles to the Moon, the effect is minimized, resulting in a low *neap* tide.

The effect of tides in the open oceans is negligible, perhaps one metre, and enclosed areas of water such as the Black Sea exhibit differences in the order of centimetres. However, in

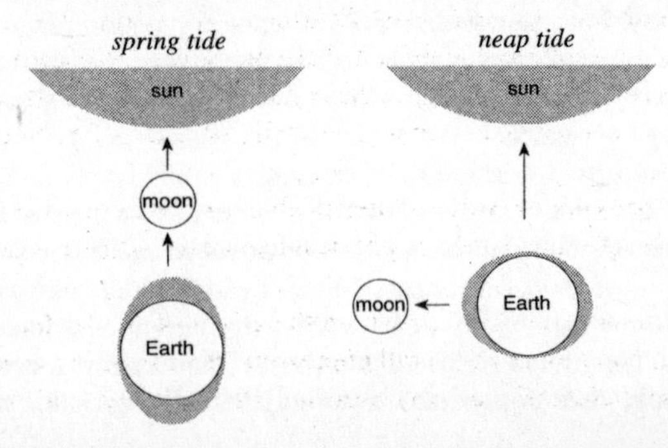

shallow seas where the tide may be channelled by the shores, tides of six to nine metres may be created.

The tides around Britain's coasts vary markedly, in part because of the effect of the *Coriolis force*. This is when air or water is pushed to the side because of the Earth's rotation. Hence, in the northern hemisphere, water moving across the surface is pushed to the right (and conversely in the southern hemisphere). Therefore the tidal wave that passes northwards up the Irish Sea creates higher tides on the Welsh and English coasts than on the Irish side, and as the tidal wave moves into the North Sea, the Coriolis force pushes the water to the right, giving higher tides on the British coastline than on the coasts of Norway and Denmark. The potential for generating energy from tides has long been realized, and the first tidal power station was built in France and made operational in 1966.

till sediment deposited as a result of the action of glacial ice without water as an agent. The size varies from CLAY particles to rock fragments. There is a variety of tills, depending on their method of release. For example, subglacial melting gives *lodgement till*, and the thawing of stationary ice produces *melt-out till*.

time zones *see* STANDARD TIME; INTERNATIONAL DATE LINE.

tombola a sand or shingle SPIT that connects an island to the mainland. It may have been the result of LONGSHORE DRIFT depositing beach material or it may be all that is left of an eroded ISTHMUS.

topography the surface features of the Earth, whether natural or artificial. It includes the physical features, soils and vegetation.

topological map a map drawn up to show a particular feature rather than to give an overall impression of an area. For example, a topological map may show only the railways in a country. Towns and cities are shown as dots and routes as straight lines. Distance, scale and relative orientation are not particularly important; clear, uncluttered presentation is.

topology the study of a particular locality.

topsoil the upper, more fertile layers of SOIL above the SUBSOIL.

tor a small hill or mass of ROCK rising abruptly from the smoother hill slopes surrounding it. It is composed of a pile of huge, well-jointed blocks on a platform of solid rock. Tors may consist of any hard rock but are often granite. They are a feature of southwest England.

tornado a narrow column of air that rotates rapidly and leaves total devastation in its path. It develops around a centre of very low pressure with high velocity winds (well over 300 kilometres an hour) blowing anticlockwise and with a violent downdraught. The typical appearance is of a funnel or snake-like column filled with CLOUD and usually no more than 150 metres (500 feet) across.

The precise way in which a tornado is formed is not known, although it involves the interface of warm moist air with dry cooler air and an inversion of temperatures with some event acting as a trigger, possibly an intense cold front. These conditions are found in many countries at mid to low latitudes but

a tornado column

particularly so in the mid-west of the USA. The destruction created by tornadoes is partly the result of the violent winds and partly the very low pressure. This has the effect of causing buildings to explode outwards because the pressure outside exceeds that inside, and although a tornado may affect an area only 100–150 metres (325–500 feet) across, the destruction is total. A tornado is often unpredictable in its behaviour and can lose contact with the ground or retrace its routes. When it moves out over the sea, and once the tunnel has joined with the waves, a WATERSPOUT is formed.

trade gap the difference in value between imports and exports when imports are the greater.

trade winds steadily blowing easterly winds that blow from subtropical high pressure areas between latitudes 30° north and south. The winds are generally northeasterly in the northern hemisphere and southeasterly in the southern hemisphere.

the trade winds

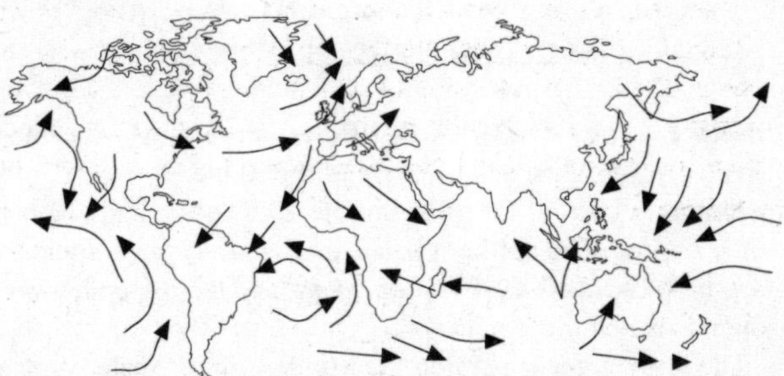

transect a transverse section across a region to show relationships between soils, vegetation and the character of the land surface.

transhumance the seasonal movement of people and animals between winter and summer pastures, e.g. to use an ALP. People have two abodes and, therefore, are not NOMADS.

tree line the limit beyond which trees do not grow. It depends on LATITUDE, ALTITUDE, soil conditions, rainfall and exposure to wind. The tree line on the sunny side of a valley is usually higher than on the shaded side.

triangulation the method used for carrying out a topographical SURVEY. A baseline is measured accurately and taken as one side of a triangle. The angles from each end of the baseline to the object to be surveyed are measured using a THEODOLITE. The lengths of the other sides of the triangle are found using trigonometry.

the position of D, e.g. a mountain top, can be calculated from the known base line AB and the measured angles, ∂ and ß

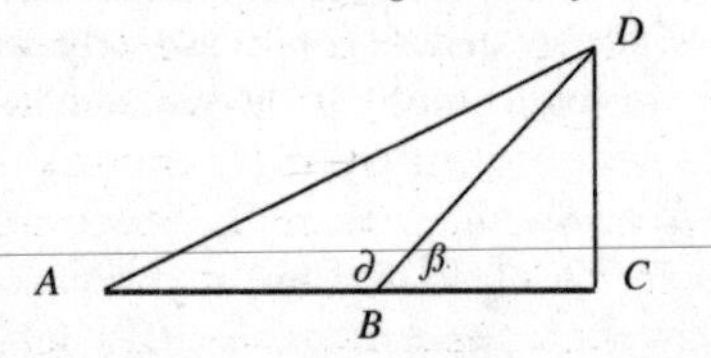

tributary a stream or RIVER that flows into a larger stream or river, e.g. the Main is a tributary of the Rhine.

tropics two lines of LATITUDE that lie 23.5° north and south of the EQUATOR. The northern line is the *Tropic of Cancer* and the southern one the *Tropic of Capricorn*, and the region between them is called the tropics.

The term *tropical* is often used to describe climate, vegetation, etc, but it is not an accurate usage of the word. In general, a tropical climate does not have a cool season and the mean temperature never falls below 20°C. Rainfall can be very high indeed, and in many countries these conditions produce a dense, lush vegetation, e.g. TROPICAL RAINFORESTS.

tropical air mass an air mass formed in the subtropical ANTICYCLONE belt. It may form over oceans, when it is given the symbol mT (maritime Tropical), or it may form over a continental interior, symbol cT (continental Tropical). It brings warmer weather as it moves away from the EQUATOR, either northwards or southwards, and is the principal means of transferring heat away from the Equator.

tropical climate a CLIMATE in which there is no cool season. Mean monthly temperatures are always above 20°C. Within the TROPICS, the climate varies according to ALTITUDE and rainfall.

tropical cyclone a low pressure system with very strong winds occurring within the TROPICS. *See* CYCLONE; HURRICANE.

tropical rainforest a forest occurring in tropical areas of heavy rainfall. Trees have long, straight trunks and are usually about 30 metres high. The top 'layer' of leaves is known as the *canopy*. There are two distinct layers lower down, each layer supporting different wildlife. There is little undergrowth. However, the variety of plant, bird and other animal life at all levels of the forest is immense. Trees shed their leaves at random because there are no marked seasons.

Tropical forests supply sought-after hardwoods, such as teak, iroko and some of the mahoganies, for the furniture industry. Indiscriminate large-scale felling leads to soil EROSION and upsets the ecology of the area. Various plants and animals may become extinct as their habitat is disturbed. The tropical rainforest is also a potential source of medicines, which can be manufactured from tree bark, etc.

It is believed that wholesale felling would have a serious effect on the world's climate as the rainforest absorbs vast quantities of carbon dioxide and gives out oxygen during photosynthesis.

troposphere the part of the Earth's atmosphere between the surface and the *tropopause* (the boundary with the STRATOSPHERE). The tropopause is the point at which the change in temperature with height (the *lapse-rate*) stops and the temperature remains constant for several kilometres. Within the troposphere itself, the temperature decreases approximately 6.5°C (12°F) for each kilometre (3280 feet) of height. The troposphere is also the layer that contains most of the water vapour and about 75 per cent of the weight of gas in the atmosphere, and it is the zone where turbulence is greatest and most weather features occur. The level of the tropopause, and therefore the top of the troposphere, varies from about 17 kilometres (55,770 feet) at the EQUATOR, falling to 9 kilometres (29,500 feet) or lower at the poles. The height variation relates to temperature and pressure at sea level.

trough (1) the lowest part of a wave between two crests. (2) an elongated area of low ATMOSPHERIC PRESSURE. All fronts occupy troughs but the reverse does not apply. (3) a U-shaped valley that has been gouged out by glacial EROSION.

true north and true south the direction of the geographic north or south pole from the observer. It differs from the direction of the magnetic poles.

tsunami (*plural* **tsunami**) an enormous sea wave caused by the sudden large-scale movement of the sea floor, resulting in the displacement of large volumes of water. The cause may be an EARTHQUAKE, volcanic eruption, a submarine slide or slump of sediment, which may itself have been started by an earthquake or tremor. The slipping of thousands of tonnes of rock from the sides of FJORDS may also cause tsunami.

The effect of this sea floor movement in the open ocean may not be seen at all, as the resulting wave may only be one metre or less in height. However, because the whole depth of water is affected, there is a vast amount of energy involved, so when the waves reach shallow water or small bays, the effects can be catastrophic. The waves may travel at several hundred kilometres an hour (600–900 or 375–560 miles) and reach heights of 15–30 metres (50–100 feet). The devastation likely to be caused is clearly great, and there are many instances on record.

The word originates from Japanese (*tsu*, 'harbour', and *nami*, 'waves'), and in Japan there have been many instances of destructive tsunami. In 1933, an earthquake triggered one with waves up to 27 metres (88 feet) high and thousands of people were drowned along the Japanese coast. The waves were recorded about 10 hours later in San Francisco, having crossed the Pacific. It seems that tsunami are generated by submarine earthquakes registering 8 or over on the RICHTER SCALE.

tufa a SEDIMENTARY ROCK, usually composed of calcium carbonate precipitated out of solution at springs, where there may be heating, and expulsion of carbon dioxide (i.e. the loss of CO_2, which causes the deposition of $CaCO_3$). Tufa tends to show a porous (almost spongy) form and is often interbedded with sands or gravels.

tundra the treeless region between the snow and ice of the ARCTIC and the northern extent of tree growth. Large treeless plains can be found in northern Canada, Alaska, northern Siberia and northern Scandinavia. The ground is subject to PERMAFROST, but the surface layer melts in the summer, so soil conditions are very poor, being waterlogged and MARSHY. The surface therefore can support little plant life. Cold temperatures and high winds also limit the diversity of plants, restricting the *flora* to grasses, mosses, lichens, sedges and dwarf shrubs. Some areas of tundra receive the same low level of PRECIPITATION as DESERTS, yet the soil remains saturated as a result of the partial thaw of the permafrost. Most growth occurs in rapid bursts during the almost continuous daylight of the very short summers.

Because of the inhospitable conditions, animal life is also limited, although more numerous in summer. In addition to insects (midges, mosquitoes, etc) and migratory birds, there are wolves, Arctic foxes, lemmings, hares, snowy owls and the herbivorous reindeer in Europe and caribou in northern America. Polar bears occur at the coast.

In addition to this Arctic tundra, there is also *alpine tundra*, which is found on the highest mountain tops and is therefore widely spread. However, conditions differ because of daylight throughout the year and plant growth in the tropical alpine tundra also occurs all year round.

turbidite a SEDIMENTARY ROCK type, deposited by a TURBIDITY CURRENT, which, because of the nature of the current, is variable in thickness and extent. Turbidites are a major sediment type derived from the land and deposited on continental margins. The deposits often show a well-defined and characteristic internal division of graded sands at the base, followed by laminated sands, then laminated silty sands with ripple marks. The uppermost layers comprise fine-grained silts and muds. It is thus, overall, a fining-upwards cycle, i.e. the grain size of the sediments decreases upwards.

turbidity current a current in water that is loaded with sediment, producing a gravity-controlled body of water and sediment (within a sea, lake, etc) that is denser than the surrounding water. Sediments on a slope within the sea, lake or DELTA are disturbed, perhaps by earthquake tremors, and the current flows downslope. The flow moves rapidly on the floor at speeds up to 7 metres (23 feet) per second, and sediment is deposited at the foot of the slope or on the near-level, deep ocean floor. The currents are usually short-lived and have the effect of depositing shallow water sediments in deep water environments. The sediments thus formed are called TURBIDITES. In carrying suspended sediment at relatively high speeds, turbidity currents possess considerable erosional force. This has been demonstrated on numerous occasions, and none better than the flows triggered by the Grand Banks earthquake (off Newfoundland) in 1929, where the current broke through numerous submarine cables.

twilight the period of partial light after sunset or before sunrise, caused by the reflected sunlight in the upper ATMOSPHERE when the Sun is below the horizon. In astronomy, it is defined as beginning when the Sun is 18° below the horizon.

twilight area a rundown area in a city. Housing and facilities are of a very poor standard, and the level of crime and vandalism is likely to be high (*see* URBAN DECAY).

twister a colloquial term used in the USA to describe a WATERSPOUT or a TORNADO.

typhoon *see* HURRICANE.

U

ubac a French term to describe the side of a VALLEY that is shaded, i.e. in the northern hemisphere, it is the north-facing side. It is less likely to be used for growing crops or for housing.

ultraviolet radiation a form of radiation that occurs beyond the violet end of the visible light spectrum of electromagnetic waves. Ultraviolet rays have a frequency ranging from 10^{15}Hz to 10^{18}Hz, with a wavelength ranging from 10^{-7} metres to 10^{-10} metres. They are part of natural sunlight and are also emitted by white-hot objects (as opposed to red-hot objects, which emit infrared radiation). As well as affecting photographic film and causing certain minerals to fluoresce, ultraviolet radiation will rapidly destroy bacteria. Although ultraviolet rays in sunlight will convert steroids in human skin to vitamin D (essential for healthy bone growth), an excess can cause irreversible damage to the skin and eyes and damage the structure of the DNA in cells. Fortunately, a great deal of the ultraviolet radiation from the Sun does not reach the Earth as the OZONE LAYER in the upper ATMOSPHERE acts as a filter.

underfit stream *or* **misfit stream** a stream that has narrower MEANDER belts and shorter meander wavelengths than would be expected in the VALLEY in which it flows. It is possible that the valley was formed by a larger and faster flowing river than the present stream. Underfit streams may be evidence of a change in the climate.

undertow a strong, seaward CURRENT that occurs where waves break on a beach. After the SWASH has pushed water up the beach and it has run back down with the BACKWASH, the undertow carries that water back out to sea along the sea bed. It is usually more marked alongside an obstruction than in the middle of an open beach. It may be the initial stage of a RIP CURRENT.

uniform delivered pricing a system practised in Britain by the major supermarkets and chain stores. The cost of an item will be the same, regardless of the distance of the retail outlets from the central warehouses. For example, a bag of sugar from a branch of a supermarket in the north of Scotland costs the same as in the Midlands.

updraught a rising current of air, associated with THUNDERSTORMS and CUMULONIMBUS CLOUDS. Powerful updraughts are at the centre of all WHIRLWINDS and are a feature of all unstable AIR

MASSES. They enable glider pilots and gliding birds to gain height. *See* CONVECTION; DUST DEVIL; TORNADO; WATERSPOUT.

upslope fog an ADVECTION FOG that occurs in upland areas. Rising ground pushes moist air upwards. As the air rises, it is cooled until the temperature falls below saturation point and water vapour condenses out. As the fog drifts upslope, it is replaced by more cooling air, which, in turn, forms more fog.

upwelling the means by which cold, deep sea CURRENTS rise through the OCEAN to the warmer surface layers. The cold currents are rich in nutrients, which enable plankton to grow and thus provide feed for fish shoals. The world's best fishing grounds are usually found where there is an upwelling current, e.g. BENGUELA, HUMBOLDT CURRENTS. The upwelling of cold currents also plays a part in the circulation of ocean currents.

urban pertaining to a city or town. There are no limits of size or population, so an *urban area* in one country may not be regarded as an urban area in another. For example, a settlement of 300 people is regarded as an urban area in Iceland, whereas 10,000 or more people are required by other countries in order for a settlement to be classed as urban.

urban decay the older, rundown part of a city where housing has become dilapidated and vandalism and petty crime are common. Urban decay has always been a problem. In postwar years the substandard, inadequate housing was known as *slums*, and programmes of slum clearance took place, with inhabitants being relocated to new council housing on the outskirts of cities. In the 1990s, various government initiatives are underway in an attempt to redevelop rundown city centres. These have been designated *urban development areas*, and there are various incentives, such as relaxed planning conditions, to encourage developers. *See* TWILIGHT AREA.

urban diseconomies the disadvantages of urban life, e.g. traffic congestion, pollution, higher rates and higher house insurance.

U-shaped valley *or* **glaciated valley** the GLACIER occupied the whole VALLEY, rather than just the valley floor, and its action over-steepened the sides of the valley and widened it. Irregularities on the valley floor are smoothed out, although bars of highly resistant rock may remain or PATERNOSTER LAKES may have been formed. There may be associated HANGING VALLEYS.

V

vadose pertaining to the layer above the surface of the WATER TABLE and immediately below the surface of the ground. Water content is variable both in amount and position. Where the ROCK is permeable, water can percolate freely through the soil under the influence of gravity, particularly in limestone areas.

valley an elongated depression in the land surface. It is created initially by river EROSION (V-SHAPED VALLEY) but may be modified later by glacial erosion to form a U-SHAPED VALLEY. *See also* HANGING VALLEY; RIFT VALLEY.

valley glacier a GLACIER situated in an upland preglacial valley. The valley glacier may be the result of the coalescence of several CIRQUE glaciers or it may occur at the edge of an ICE SHEET or ICE CAP.

valley train a deposit of SAND, gravel and pebbles stretching along the sides of a VALLEY for a considerable distance in front of the snout of a GLACIER. Many valley trains have been cut by FLUVIAL EROSION, leaving the remains as terraces.

valley wind the wind generated in a VALLEY as air is warmed by the Sun during the day and cooled at night. The warmed air forms an upslope, ANABATIC wind during the day, and there is a corresponding downslope cool KATABATIC wind during the night. The effect is most marked in east-west orientated valleys, as the south-facing slope collects more sunshine and there is a marked temperature difference between it and the north-facing slope.

Van Allen belts charged particles that are trapped in the Earth's magnetic field and form two belts around the Earth. The lower belt occurs between about 2000 and 5000 kilometres (1240 and 3100 miles), and its particles are derived from the Earth's ATMOSPHERE. The particles in the upper belt are from the SOLAR WIND and the belt occurs at around 20,000 kilometres (about 12,400 miles). The belts were named after the American space scientist James Van Allen (1914–), who discovered them in 1958.

vapour pressure that part of the ATMOSPHERIC PRESSURE that is due to water molecules contained in the atmosphere. Maximum vapour pressure at a given temperature occurs when the air is saturated. It varies from about 10 millibars in the UK to about 30 millibars near the EQUATOR.

varve a lacustrine deposit, near to ICE SHEETS, of banded CLAYS, silts and sands. The rhythmically banded sediments were deposited annually in lakes at the edge of ice sheets. Spring meltwaters bring new loads of sediment into the lake; the coarse particles are deposited quickly while the finer particles are only deposited from suspension later in the year. Since the glacial streams would refreeze in the winter, the sediment supply would cease until the following spring. This cyclic activity produces the banded effect, and each season accounts for one pale coarse band and one dark finer band. Varve deposits are very thin, and 50 or 70 years may be accommodated in one metre of sediment.

veering a term used to describe the wind as it moves in a clockwise direction, i.e. in the northern hemisphere from north to east and vice versa in the southern hemisphere. *See* BACKING.

vein in geology, a sheet-like feature, usually occupying a fracture or fissure within a ROCK, that is infilled with mineral deposits. Calcite and quartz commonly form veins, but ore deposits do occur in this form, commonly mixed with other minerals.

veld *or* **veldt** the Afrikaans term for the naturally occurring open grasslands of South Africa. It is further classified according to ALTITUDE, i.e. the *high veld*, *middle veld* and *low veld*, and to the type of underlying soil: *sand veld*, *grass veld*, BUSH VELD. *See also* PRAIRIE; PAMPAS; STEPPES.

vent an opening in a VOLCANO leading down to the MAGMA chamber through which LAVA is ejected during an eruption. Some volcanoes have a single central vent, others have a line of vents.

Venturi effect the effect that a narrowing channel has on the velocity of a current of gas or liquid. It is particularly marked as wind blows down a narrow VALLEY or between high buildings. The velocity of the wind is increased significantly in both cases and should be taken into account when high buildings are being designed.

virga a feature relating to certain CLOUD formations, where trails of PRECIPITATION (rain or snow) fall beneath the cloud but evaporate before reaching the ground.

volcano a natural vent or opening in the Earth's crust that is connected by a pipe, or *conduit*, to a chamber at a depth that contains MAGMA. Through this pipe (usually called a VENT) may be ejected LAVA, volcanic gases, steam and ash, and it is the amount of gas held in the lava and the way in which it is released on reaching the surface that determine the type of eruption. Volcanoes may be *active*, i.e. actually erupting whether just clouds of ash and steam or lava; *extinct*, i.e. the activity ceased a long time ago; or *dormant*. Dormant volcanoes have often in the past been thought to be extinct, only to erupt again with startling ferocity.

Volcanoes can be described by the type of eruption, which is named after a particular volcano that exhibits a specific eruption pattern:

Hawaiian	outpouring of fluid lava and little explosive activity
Peléean	violent eruptions with viscous lava and *nuées ardentes**
Strombolian	moderate eruptions, small explosions and lava of average viscosity
Vesuvian	very explosive after a dormant period with ash/gas clouds and gas-filled lava
Plinian	very explosive with *pyroclastics*† ejected in a column up to 50 kg high producing thick airfall deposits

* nuées ardentes—an old term meaning an incandescent ash flow that moves rapidly

† pyroclastics—volcanic rocks formed from broken fragments, e.g. bombs, pumice, ash, cinders

Volcanoes of the Hawaiian type are also called *shield volcanoes*. The sides of the volcanoes are almost flat because of the rapid flow of the lava. *Composite volcanoes* show greater angles of slope due to a build-up of lava and pyroclastic material. Both shield and composite types are also called *central type*, because the supply comes from a central vent, as opposed to *fissure volcanoes*, which erupt through splits where the crust is under tension.

Active volcanoes occur in belts associated with the tectonic plates (*see* PLATE TECTONICS) with about 80 per cent of the active subaerial volcanoes at destructive plate margins, 15 per cent at constructive plate margins and the remainder within plates. Most submarine volcanism is at constructive plate margins.

The environmental effect of volcanoes can be very significant, whether it be the enormous amounts of ash ejected into the atmosphere or the consequences of lava flows consuming the countryside. At the time of eruption, volcanic materials are often over 1000°C, hence flows either burn, push over or cover whatever they meet.

Over 500 volcanoes have been active in historic times but only about 50 erupt each year, often on a very small scale.

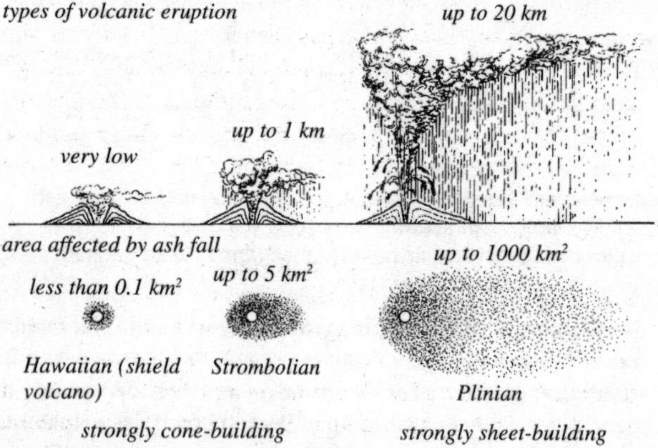

types of volcanic eruption

up to 20 km

up to 1 km

very low

area affected by ash fall

up to 1000 km²

less than 0.1 km²

up to 5 km²

Hawaiian (shield volcano)

Strombolian

Plinian

strongly cone-building

strongly sheet-building

V-shaped valley a valley that has been created by RIVER EROSION. It has evenly sloping sides, in contrast to U-SHAPED or glaciated valleys, which have much steeper sides. The upper reaches of most unglaciated river valleys are V-shaped, although shape is influenced by other factors. In humid climates, the V shape is widened by erosion, while in dry climates the V may be steep and narrow. Where a valley has a sunny and a shaded slope, the V is likely to be uneven as vegetation and erosion are likely to be different on each slope. The ROCK structure of the valley will also affect shape. Valleys cut through CLAY tend to be deep and steep. The V shape tends to be lost near the mouth of a river where the plain broadens out.

W

wadi a usually dry VALLEY that has been formed by RIVER action. It occurs in ARID or semi-arid areas and may be half-filled with sand. It may carry water after a storm, and in such cases there is usually considerable EROSION. Because of the dry climate, there is little weathering effect and the sides of the wadi tend to be steep and GORGE-like.

waning slope the low, slightly concave slope at the foot of a hill. It is produced by the action of water carrying surface soil down from the slope above or by the retreat of the slope. It has been claimed that the waning slope grows in area as erosion proceeds. *See* PEDIMENT.

warm front the edge of a mass of warm air advancing and rising over cold air. CLOUD develops in the rising air, with heavy NIMBOSTRATUS forming as the FRONT passes. There is usually heavy rain associated with the front. The temperature rises after the front, with the rain clearing and often a change in wind direction.

warp the very fine-grained sediments found in tidal ESTUARIES. The Soil Survey of England and Wales describes alluvial deposits as warp soils.

wash (1) the name of a large area of tidal sandbanks on the east coast of England, viz. the Wash. (2) the name given to coarse-textured alluvial material. (3) the movement downslope of surface soil by rain.

washland a FLOOD PLAIN with artificial embankments into which a RIVER can be diverted in order to prevent flooding and damage further downstream where there are settlements.

water balance a term used in meteorology to describe the movement of water, on a global scale, between the ATMOSPHERE and the ground surface. The water balance is concerned with all forms of PRECIPITATION, EVAPORATION, HUMIDITY and the flow of water in RIVERS and streams. Information on OCEAN CURRENTS and sea temperatures are also included, but, since this information is relatively scarce, the accuracy of the water balance over the oceans is not reliable. It is more reliable over continental areas.

waterfall a point in the course of a RIVER where water falls vertically. The position of a waterfall will move very slowly upstream as the lip of the fall is eroded. The pool at the bottom of the fall is also eroded by the force of the water as it plunges down. Waterfalls may be the result of descent from the edge of a PLATEAU, resistant ROCK bands across the stream of the river, descent from a HANGING VALLEY into a glaciated TROUGH, or descent from a cliff into the sea. Niagara Falls and Victoria Falls are among the most spectacular waterfalls in the world.

water meadow a flat area beside a river that is subject to flooding from time to time, usually in winter. The flood may deposit a layer of sediment, which encourages growth the next spring. Water meadows provide good grazing for farm animals.

watershed the boundary between two DRAINAGE BASINS. It usually runs along the highest ground where the head streams of RIVERS rise.

waterspout the water equivalent of a TORNADO, although it is shorter-lived and of lower intensity. Water is sucked up by the UPDRAUGHT from the surface of the LAKE or sea and added to water that condenses out of the swirling air. When the waterspout collapses, the water falls as an immense deluge that is much more noticeable if it happens on land. The waterspout is caused by intense local heating. As the warmed air rises a small, local low pressure system is formed along with associated CUMULONIMBUS CLOUDS.

water table the level below which water saturates the spaces in the ground; the top of the zone where groundwater saturates permeable ROCKS. It is where ATMOSPHERIC PRESSURE is equalled by the pressure in the groundwater. The position (*elevation*) of the water table varies with the amount of rainfall, etc, loss through evaporation and transpiration from vegetation, percolations through the soil. A SPRING or seepage occurs when, as a result of geological conditions, the water table rises above ground level (*see also* AQUIFER).

wave (1) a crest of water between two troughs. Waves are set up by the pressure of the wind across the water surface. The height of the wave and its wave length (the distance between succeeding crests) is in direct proportion to the strength of the wind when there is no interference from the sea bed. As waves approach the shore, wave length decreases and height increases until eventually the wave breaks onto the shore. The energy of a breaking wave is responsible for much coastal EROSION, but waves are also responsible for depositing material on beaches. *See* LONGSHORE DRIFT. Some waves are caused by underwater EARTHQUAKES (*see* TSUNAMI) or by landslides into the sea or a LAKE. *See* BACKWASH; BREAKER; CONSTRUCTIVE WAVE; DESTRUCTIVE WAVE; SURF; SWASH; SWELL. (2) atmospheric waves, which include ROSSBY waves. (3) electromagnetic waves that carry sound, light and heat and include radiowaves, microwaves, X-rays and gamma rays.

wave base water particles in a WAVE move in circles as the wave moves forward. The diameter of these circles of orbital motion decreases with depth of the sea. Wave base is the depth at which there is no motion caused by the wave at the surface. At this level, sediments are not disturbed by wave action at the surface.

wave base

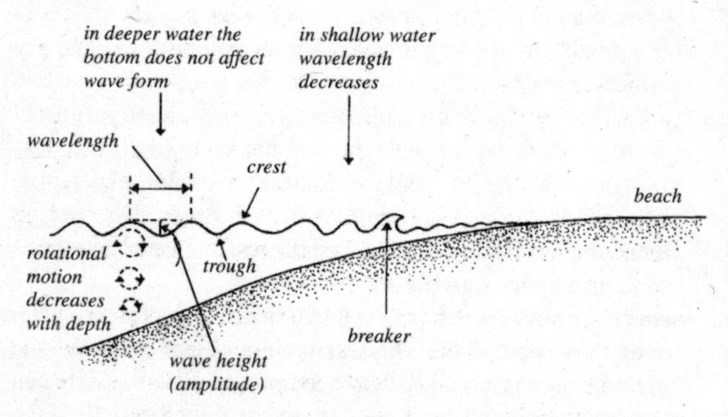

wave refraction the process whereby WAVES bend so that they are parallel to the shore when they break onto it. The velocity of a wave decreases as the depth of water decreases, i.e. the wave slows down in shallow water. This means that the part of the wave in the centre of a bay where the water is deeper will run on more quickly than the parts over shallower water at the sides of the bay. In other words, the wave curves or becomes refracted.

Wave refraction also occurs as waves approach a straight beach at an acute angle. The end of the wave nearest the shore

weather because it is at a location where many different AIR MASSES may meet. The surface at which two air masses with

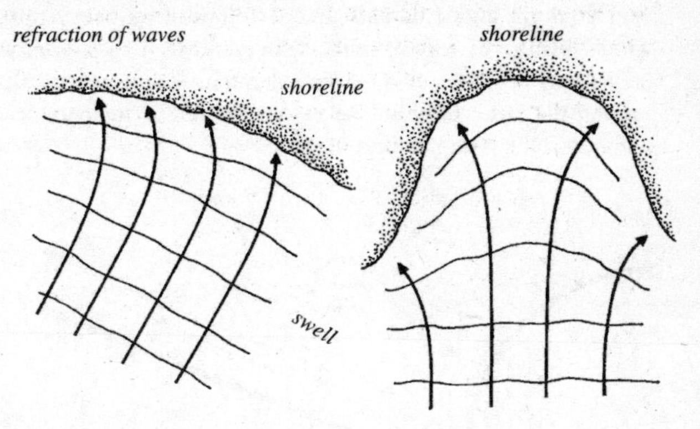

encounters shallow water before the rest of the wave and is refracted towards the shore.

wealth-consuming sector the SERVICE INDUSTRIES. 'New' wealth is not created. The service industries are able to exist when a community can earn enough money to pay for them. *See* TERTIARY INDUSTRY.

wealth-creating sector manufacturing industry. Raw materials are refined and made into items that can be sold at a profit. That profit can be reinvested in manufacturing industry, used to purchase manufactured goods or spent in the service sector.

weather the combined effect of ATMOSPHERIC PRESSURE, temperature, sunshine, CLOUD, humidity, wind and the amount of PRECIPITATION that together make up the weather for a certain place over a particular (usually short) time period. The weather varies enormously around the Earth, but some countries have stable weather patterns while Britain tends to have changeable

different meteorological properties meet is called a FRONT. A *warm front* occurs in a depression, between warm air moving over cold air, and it heralds drizzle followed by heavy rain, which then gives way to rising temperatures. A *cold front* is the leading edge of a cold air mass that moves under warm air, forcing the latter to rise. The result is a fall in temperature, with rainfall passing behind the front.

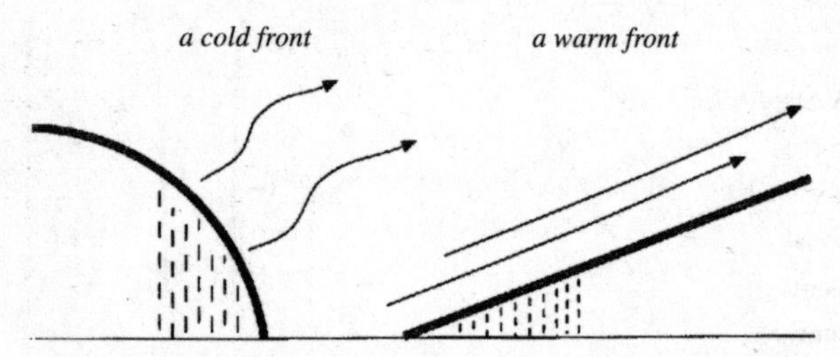

On a weather map (not those seen on television), the components of the weather are represented by symbols (*see* pages 183–184), which enable meteorologists to summarize a lot of information early.

CLIMATE is weather taken over a long time period, 30 years or more.

weathering a combination of chemical and physical processes on the surface of the Earth, or very near to it, which breaks down rocks and minerals. Weathering takes various forms and can be divided into *mechanical*, *chemical* and *organic*.

Some types of weathering

mechanical	*freeze-thaw action* (alternate freezing and thawing of water in cracks producing widening or break-up)
	exfoliation peeling off in thin rock layers (like onion skin)
	disintegration into grains

Weather symbols

	Beaufort letter(s)	*Symbol*
State of the sky		
fine—blue sky	b	○
fair—sky partly cloudy	bc	⦶
cloudy—sky very cloudy	c	◍
overcast—sky completely covered	o	●
Precipitation		
drizzle	d	,
rain	r	•
snow	s	✱
sleet	rs	✱̇
showers		
—passing, of rain	pr	∇̇
—passing, of snow	ps	✱∇
—passing, of hail	ph	△∇
thunderstorms		
—with rain	tlr	⌈Κ•
—with snow	tls	⌈Κ✱
—with hail	tlh	⌈Κ△
hail	h	△
thunder	t	T
lightning	l	⟨
Visibility		
fog		
—over the sea (coast station)	fs	≡
—over low ground (inland)	fg	=

Weather symbols contd

	Beaufort letter(s)	*Symbol*
mist—i.e. visibility reduced to $^1/_2$–1 mile	m	—
haze—visibility reduced by dust	z	∞
Other phenomena		
dew	w	⌓
hoar frost—dew as ice crystals	x	⊔
squalls—a line of violent showers	kq	∀
storm of drifting snow	ks	✛
sandstorm or dust storm	kz	S

chemical *carbonation* the reaction of weak carbonic acid (H_2CO_3) with the rock
hydrolysis combination of water with minerals to form insoluble residues (e.g. clay, minerals)
oxidation and *reduction*

organic *breakdown* by flora and fauna, e.g. burrowing animals, tree roots, and the release of organic acids from decomposed plants which react with minerals

These weathering processes together produce a layer of material that may then be moved by processes of EROSION.

weir a small dam built across a RIVER to maintain a suitable level of water upstream. This may be for navigation, in which case ships would bypass the weir by using locks. Weirs may be built to maintain a suitable level in salmon fishing pools. In this case, the weir will also act as a salmon ladder. There will always be a sluice or notch in a weir to enable the excess river water to escape downstream.

westerlies the winds that predominate between 40° and 70° north and south of the EQUATOR. The airflow is from the SUB-

TROPICAL high pressure areas to the low pressure areas of the temperate zone. Prevailing winds are southwesterly in the northern hemisphere and northwesterly in the southern hemisphere. The vast expanses of OCEAN, unbroken by land masses in the southern hemisphere, mean that stationary pressure systems do not develop. The result is that, in general, westerly winds in the southern hemisphere are stronger than westerlies in the northern hemisphere (*see* ROARING FORTIES).

west wind drift a slow movement of OCEAN water eastwards. In the northern hemisphere, it occurs between latitudes 35° and 45° in the Pacific and is known as the *North Pacific current*. In the North Atlantic it is known as the GULF STREAM, which starts in the Caribbean and flows diagonally northeastwards. As it approaches Britain, it is called the *North Atlantic drift*. In the southern hemisphere, it is a continuous current flowing right round the globe. In the Pacific it is a relatively warm current at 45°–60° south, while in the Atlantic it is cooler and is found between 40° and 65° south.

whirlwind a rapidly rotating column of air. It is caused by local surface heating leading to strong CONVECTION CURRENTS. As the hot air rises, a small, local, low pressure system is created, around which the whirlwind revolves. It is usually a short-lived phenomenon. *See* DUST DEVIL; TORNADO; WATERSPOUT.

WHO (abbreviation of World Health Organization) an international organization founded in 1948 to develop good health and to promote the provision of health care, especially in developing countries.

white-out a state of the ATMOSPHERE in which normal perception of distance and shape is lost and snow-covered ground merges into the sky without a visible horizon. It tends to occur during or after a BLIZZARD, when multiple reflection from CLOUD and snow crystals prevent shadows from forming. It is common in polar regions but can also occur in MOUNTAIN areas. It is very disorientating.

wind a generally horizontal or near-horizontal movement of air caused by changes in ATMOSPHERIC PRESSURE in which air normally moves from areas of high to low pressure. Wind speed is greater when the ISOBARS (lines joining points at the same pressure) are closely packed on weather maps, and the BEAUFORT SCALE provides a systematic guide to wind speed. Because of the Earth's rotation and the effect of the Coriolis force (*see* TIDES), air in the northern hemisphere flows clockwise around a high pressure area and anticlockwise around a low.

The TRADE WINDS play an important part in the atmospheric circulation of the Earth, and they are mainly easterly winds that blow from the subtropics to the EQUATOR. The westerlies flow from the high pressure of the subtropics to the low pressure of the temperate zone. The WESTERLIES form one of the strongest wind flows and their strength increases with height (*see* JET STREAM), and DEPRESSIONS are most common in this wind system. The DOLDRUMS is a zone of calms or light winds around the Equator, applied particularly to the oceans, with obvious links to the time when sailing ships were becalmed. Also linked to sailing are the ROARING FORTIES, which are westerlies in the southern hemisphere where they tend to be stronger. However, the supposed link of trade winds with early travel on the sea is incorrect—their origin is from the Latin word meaning 'constant'.

In addition to its destructive power, wind provides an additional hazard when combined with cold. *Wind chill* is the effect wind has in lowering apparent temperatures through increasing heat loss from the body. For example, in calm conditions at –12°C (10°F) there is little danger for someone properly clothed, but if the wind speed is 40 kilometres (25 miles) per hour, then the wind chill creates an equivalent temperature of –34°C (–29°F), which is potentially harmful.

windbreak any barrier that reduces wind speeds at the surface of the ground. Trees are frequently planted to provide a windbreak or shelter belt. Hedges provide windbreaks, and their removal can lead to EROSION of the soil. *See* DUST BOWL.

wind rose a diagram to illustrate the frequency with which WIND blows from various specified directions in a given period of time. The number of calm days is given in a circle at the centre of the rose. The lengths of the arms represent the number of days of wind from each of the chosen directions. The arms can be subdivided to show the frequency of different wind speeds.

example of a wind rose

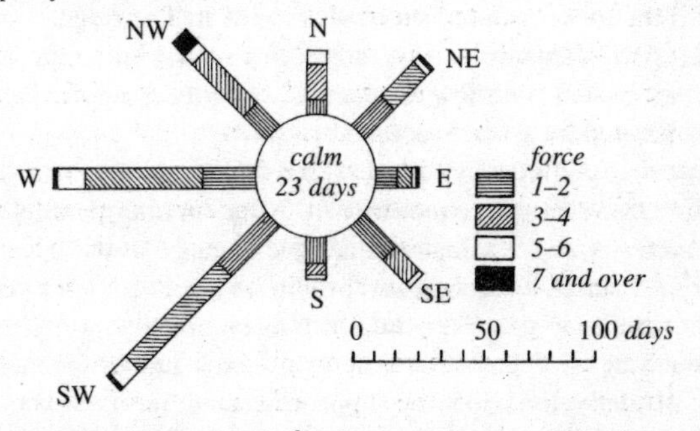

wind shadow a zone of quiet air in the LEE of an object or WINDBREAK. Although there is shelter from the full force of the WIND, it is likely that EDDIES will be felt and, in times of SNOW, snowdrifts will form in areas of wind shadow.

windward the side exposed to the direction from which the wind is blowing. It is the opposite of LEEWARD.

wold an area of rolling CHALK DOWNS on which there are no trees. It is frequently used in southern England as part of a place name.

XYZ

xerophyte a species of plant that has adapted to growing in very dry conditions. The strategies to avoid water loss include the development of hard, glossy leaves; the development of very small narrow leaves; the substitution of spines and thorns for leaves; the ability to store water in thick, bulbous stems; or the growth of thick bark. Cacti are examples of xerophytes.

yardang a long ridge of ROCK, lying along the direction of the PREVAILING WIND in a DESERT. It is formed as the wind removes all the loose material on either side of it. The ridge is sharply crested. Yardangs can be only a few metres in length or they may stretch for a few kilometres. Neighbouring yardangs are separated by a wind-scoured groove.

yazoo a TRIBUTARY that is prevented from joining a main stream by LEVEES that it cannot breach. After running parallel to the main stream for some considerable distance, it will eventually find a suitable junction and be able to join the main river.

year the time taken for the Earth to complete one orbit round the Sun—365 days, 5 hours, 48 minutes and 46 seconds. The calendar year is 365 days long. To correct for the 'extra' time, every fourth year is a Leap Year and is 366 days long.

zenith the point on the celestial sphere that is vertically above the observer. The position of the zenith is important when calculating the amount of SOLAR RADIATION reaching the Earth.

zeuge (*plural* **zeugen**) a ROCK or YARDANG in a DESERT, the base of which has been undercut by WIND EROSION. The top layers of rock are harder and are eroded much more slowly than the lower layers.

Earth's Vital Statistics	
Age:	Approx 4600 million years
Weight:	Approx 5.976 x 10^{21} tonnes
Diameter:	Pole to Pole through the centre of the Earth 12,713 km (7900 miles) Across the Equator through the centre of the Earth 12,756 km (7926 miles)
Circumference:	Around the Poles 40,008 km (24,861 miles) Around the Equator 40,091 km (24,912 miles)
Area:	Land 148,326,000 sq km (57,268,700 sq miles) 29% of surface Water 361,740,000 sq km (139,667,810 sq miles) 71% of surface
Volume:	1,084,000 million cubic km (260,160 million cubic miles)
Volume of the oceans:	1321 million cubic km (317 million cubic miles)
Average height of land:	840 m (2756 ft) above sea level
Average depth of ocean:	3808 m (12 493 ft) below sea level
Density:	5.52 times water
Mean temperature:	22°C (72°F)
Length of year:	365.25 days
Length of one rotation:	23 hours 56 minutes
Mean distance from Sun:	149 600 000 km (92,960,000 miles)
Mean velocity in orbit:	29.8 km (18.5 miles) per second
Escape velocity:	11.2 km (6.96 miles) per second
Atmosphere:	Main constituents: nitrogen (78.5%), oxygen (21%)
Crust:	Main constituents: oxygen (47%), silicon (28%), aluminium (8%), iron (5%).
Known satellites:	1 (the Moon)

Continents of the World

	Highest Point		Area	
	(m)	(ft)	(sq km)	(sq miles)
Asia	8848	29,028	43,608,000	16,833,000
Africa	5895	19,340	30,335,000	11,710,000
North & Central America	6194	20,320	25,349,000	9,785,000
South America	6960	22,834	17,611,000	6,798,000
Antarctica	5140	16,863	14,000,000	5,400,000
Europe	5642	18,510	10,498,000	4,052,000
Oceania	4205	13,796	8,900,000	3,400,000

Oceans of the World

	Maximum Depth		Area	
	(m)	(ft)	(sq km)	(sq miles)
Pacific	11,033	36,198	165,384,000	63,838,000
Atlantic	8381	27,496	82,217,000	31,736,000
Indian	8047	26,401	73,481,000	28,364,000
Arctic	5450	17,880	14,056,000	5,426,000

Principal Mountains of the World

Name (location)	Height (m)	Height (ft)	Name (location)	Height (m)	Height (ft)
Everest (Asia)	8848	29,028	McKinley (N Amer)	6194	20,320
Godwin-Austen or K2 (Asia)	8611	28,250	Logan (N Amer)	5951	19,524
			Cotopaxi (S Amer)	5896	19,344
Kangchenjunga (Asia)	8586	28,170	Kilimanjaro (Africa)	5895	19,340
Makalu (Asia)	8463	27,766	Huila (S Amer)	5750	18,865
Dhaulagiri (Asia)	8167	26,795	Citlaltepi (C Amer)	5699	18,697
Nanga Parbat (Asia)	8125	26,657	Demavend (Asia)	5664	18,582
Annapurna (Asia)	8091	26,545	Elbrus (Asia)	5642	18,510
Gosainthan (Asia)	8012	26,286	St Elias (N Amer)	5489	18,008
Nanda Devi (Asia)	7816	25,643	Popocatepetl (C Amer)	5452	17,887
Kamet (Asia)	7756	25,446	Foraker (N Amer)	5304	17,400
Namcha Barwa (Asia)	7756	25,446	Ixtaccihuati (C Amer)	5286	17,342
Gurla Mandhata (Asia)	7728	25,355	Dykh Tau (Europe)	5203	17,070
Kongur (Asia)	7720	25,325	Kenya (Africa)	5200	17,058
Tirich Mir (Asia)	7691	25,230	Ararat (Asia)	5165	16,945
Minya Kanka (Asia)	7556	24,790	Vinson Massif (Antarctica)	5140	16,863
Kula Kangri (Asia)	7555	24,784	Kazbek (Europe)	5047	16,558
Muztagh Ata (Asia)	7546	24,757	Jaya (Asia)	5030	16,502
Kommunizma (Asia)	7495	24,590	Klyucheveyskava (Asia)	4750	15,584
Pobedy (Asia)	7439	24,406	Mont Blanc (Europe)	4808	15,774
Chomo Lhar (Asia)	7313	23,992	Vancouver (N Amer)	4786	15,700
Lenina (Asia)	7134	23,405	Trikora (Asia)	4750	15,584
Acongagua (S Amer)	6960	22,834	Monte Rosa (Europe)	4634	15,203
Ojos del Salado (S Amer)	6908	22,664	Ras Dashen (Africa)	4620	15,158
Tupungato (S Amer)	6801	22,310	Belukha (Asia)	4506	14,783
Huascarán (S Amer)	6769	22,205	Markham (Antarctica)	4350	14,271
Llullailaco (S Amer)	6723	22,057	Meru (Africa)	4566	14,979
Kailas (Asia)	6714	22,027	Karisimbi (Africa)	4508	14,787
Tengri Khan (Asia)	6695	21,965	Weisshorn (Europe)	4505	14,780
Sajama (S Amer)	6542	21,463	Matterhorn/Mont Cervin (Europe)	4477	14,690
Chimborazo (S Amer)	6310	20,702			

Name (location)	Height (m)	Height (ft)	Name (location)	Height (m)	Height (ft)
Whitney (N Amer)	4418	14,495	Cook (Oceania)	3753	12,313
Elbert (N Amer)	4399	14,431	Adams (N Amer)	3752	12,307
Massive Mount (N Amer)	4397	14,424	Teyde or Tenerife (Africa)	3718	12,198
Rainier or Tacoma (N Amer)	4392	14,410	Mahameru (Asia)	3676	12,060
Longs (N Amer)	4345	14,255	Assiniboine (N Amer)	3618	11,870
Elgon (Africa)	4321	14,176	Hood (N Amer)	3428	11,245
Pikes Peak (N Amer)	4301	14,110	Pico de Aneto (Europe)	3404	11,168
Finsteraarhorn (Europe)	4274	14,022	Etna (Europe)	3323	10,902
Wrangell (N Amer)	4269	14,005	St Helens (N Amer)	2950	9677
Mauna Kea (N Amer)	4205	13,796	Pulog (Asia)	2934	9626
Gannet (N Amer)	4202	13,785	Tahat (Africa)	2918	9573
Mauna Loa (N Amer)	4169	13,677	Shishaldin (N Amer)	2862	9387
Jungfrau (Europe)	4158	13,642	Roraima (S Amer)	2810	9219
Kings (N Amer)	4124	13,528	Ruapehu (Oceania)	2797	9175
Kinabalu (Asia)	4102	13,455	Katherine (N Amer)	2637	8651
Cameroon (Africa)	4095	13,435	Doi Inthanon (Asia)	2594	8510
Fridtjof Nansen (Antarctica)	4068	13,346	Galdhöpiggen (Europe)	2469	8100
Tacaná (C Amer)	4064	13,333	Parnassus (Europe)	2457	8061
Waddington (N Amer)	4042	13,262	Olympus (N Amer)	2425	7954
Yu Shan (Asia)	3997	13,113	Kosciusko (Oceania)	2230	7316
Truchas (C Amer)	3994	13,102	Harney (N Amer)	2208	7242
Wheeler (N Amer)	3981	13,058	Mitchell (N Amer)	2038	6684
Robson (N Amer)	3954	12,972	Clingmans Dome (N Amer)	2025	6642
Granite (N Amer)	3902	12,799	Washington (N Amer)	1917	6288
Borah (N Amer)	3858	12,655	Rogers (N Amer)	1807	5927
Monte Viso (Europe)	3847	12,621	Marcy (N Amer)	1629	5344
Kerinci (Asia)	3805	12,483	Cirque (N Amer)	1573	5160
Grossglockner (Europe)	3797	12,460	Pelée (C Amer)	1463	4800
Erebus (Antarctica)	3794	12,447	Ben Nevis (Europe)	1344	4409
Fujiyama (Asia)	3776	12,388	Vesuvius (Europe)	1281	4203